102
Advances in Biochemical Engineering/Biotechnology

Series Editor: T. Scheper

Advances in Biochemical Engineering/Biotechnology
Series Editor: T. Scheper

Recently Published and Forthcoming Volumes

Tissue Engineering I

Scaffold Systems for Tissue Engineering

Volume Editors: Kyongbum Lee · David Kaplan

With contributions by

C. Chan · T. A. Holland · J. Jiang · D. Kaplan
C. T. Laurencin · Z. Li · H. H. Lu · T. Matsumoto
A. G. Mikos · D. J. Mooney · P. V. Moghe · L. S. Nair
S. Patil · C. S. Ranucci · E. J. Semler · J. Velema

Advances in Biochemical Engineering/Biotechnology reviews actual trends in modern biotechnology. Its aim is to cover all aspects of this interdisciplinary technology where knowledge, methods and expertise are required for chemistry, biochemistry, micro-biology, genetics, chemical engineering and computer science. Special volumes are dedicated to selected topics which focus on new biotechnological products and new processes for their synthesis and purification. They give the state-of-the-art of a topic in a comprehensive way thus being a valuable source for the next 3–5 years. It also discusses new discoveries and applications. Special volumes are edited by well known guest editors who invite reputed authors for the review articles in their volumes.

In references *Advances in Biochemical Engineering/Biotechnology* is abbeviated *Adv Biochem Engin/Biotechnol* and is cited as a journal.

Springer WWW home page: springer.com
Visit the ABE content at springerlink.com

Library of Congress Control Number: 2006929796

ISSN 0724-6145
ISBN-10 3-540-31944-1 Springer Berlin Heidelberg New York
ISBN-13 978-3-540-31944-3 Springer Berlin Heidelberg New York
DOI 10.1007/11579328

Springer is a part of Springer Science+Business Media

springer.com

Cover design: WMXDesign GmbH, Heidelberg
Typesetting and Production: LE-TEX Jelonek, Schmidt & Vöckler GbR, Leipzig

Printed on acid-free paper   02/3141 YL – 5 4 3 2 1 0

Advances in Biochemical Engineering/Biotechnology
Also Available Electronically

For all customers who have a standing order to Advances in Biochemical Engineering/Biotechnology, we offer the electronic version via SpringerLink free of charge. Please contact your librarian who can receive a password or free access to the full articles by registering at:

springerlink.com

If you do not have a subscription, you can still view the tables of contents of the volumes and the abstract of each article by going to the SpringerLink Homepage, clicking on "Browse by Online Libraries", then "Chemical Sciences", and finally choose Advances in Biochemical Engineering/Biotechnology.

You will find information about the

– Editorial Board
– Aims and Scope
– Instructions for Authors
– Sample Contribution

at springer.com using the search function.

Attention all Users
of the "Springer Handbook of Enzymes"

Information on this handbook can be found on the internet at springeronline.com

A complete list of all enzyme entries either as an alphabetical Name Index or as the EC-Number Index is available at the above mentioned URL. You can download and print them free of charge.

A complete list of all synonyms (more than 25,000 entries) used for the enzymes is available in print form (ISBN 3-540-41830-X).

Save 15%

We recommend a standing order for the series to ensure you automatically receive all volumes and all supplements and save 15% on the list price.

Preface

It is our pleasure to present this special volume on tissue engineering in the series *Advances in Biochemical Engineering and Biotechnology*. This volume reflects the emergence of tissue engineering as a core discipline of modern biomedical engineering, and recognizes the growing synergies between the technological developments in biotechnology and biomedicine. Along this vein, the focus of this volume is to provide a biotechnology driven perspective on cell engineering fundamentals while highlighting their significance in producing functional tissues. Our aim is to present an overview of the state of the art of a selection of these technologies, punctuated with current applications in the research and development of cell-based therapies for human disease.

To prepare this volume, we have solicited contributions from leaders and experts in their respective fields, ranging from biomaterials and bioreactors to gene delivery and metabolic engineering. Particular emphasis was placed on including reviews that discuss various aspects of the biochemical processes underlying cell function, such as signaling, growth, differentiation, and communication. The reviews of research topics cover two main areas: cellular and non-cellular components and assembly; evaluation and optimization of tissue function; and integrated reactor or implant system development for research and clinical applications. Many of the reviews illustrate how biochemical engineering methods are used to produce and characterize novel materials (e.g. genetically engineered natural polymers, synthetic scaffolds with cell-type specific attachment sites or inductive factors), whose unique properties enable increased levels of control over tissue development and architecture. Other reviews discuss the role of dynamic and steady-state models and other informatics tools in designing, evaluating, and optimizing the biochemical functions of engineered tissues. Reviews that illustrate the integration of these methods and models in constructing model, implant (e.g. skin, cartilage), or ex-vivo systems (e.g. bio-artificial liver) are also included.

It is our expectation that the mutual relevance of tissue engineering and biotechnology will only increase in the coming years, as our needs for advanced healthcare products continue to grow. Already, tissue derived cells constitute important production systems for therapeutically and otherwise useful biomolecules that require specialized post-translational processing for their safety and efficacy. Biochemical engineering products, ranging from growth factors to polymer scaffolds, are used as building blocks or signal molecules at virtually every stage of engineered tissue formation. Importantly, the realization of engineered tissues as clinically useful and commercially viable products will at least in part depend on overcoming the same efficiency challenges that the biotechnology industry has been facing. In this light, we see the interface between tissue engineering and various other fields of biochemical engineering as a very exciting area for research and development with enormous potential for cross-disciplinary education. In this regard, we anticipate that this and other similar volumes will also be useful as supplementary text for students.

We extend our special thanks to all of the contributing authors as well as Springer for embarking on this project. We are especially grateful to Dr. Thomas Scheper and Ulrike Kreusel for their incredible patience and hard work as our production editors.

Medford, August 2006   Kyongbum Lee
                                                                                           David Kaplan

Contents

Contents of Volume 103

Tissue Engineering II

Volume Editors: Kyongbum Lee, David Kaplan
ISBN: 3-540-36185-5

Contents of *Advances in Polymer Science, Vol. 203*

Polymers for Regenerative Medicine

Volume Editor: Carsten Werner
ISBN: 3-540-33353-3

Adv Biochem Engin/Biotechnol (2006) 102: 1–46
DOI 10.1007/10_012
© Springer-Verlag Berlin Heidelberg 2006
Published online: 15 July 2006

Tissue Assembly Guided via Substrate Biophysics: Applications to Hepatocellular Engineering

Eric J. Semler · Colette S. Ranucci · Prabhas V. Moghe (✉)

Department of Biomedical Engineering, Department of Chemical and Biochemical
Engineering, C230 Engineering, 98 Brett Road, Piscataway, NJ 08854, USA
Moghe@rci.rutgers.edu

Abstract The biophysical nature of the cellular microenvironment, in combination with its biochemical properties, can critically modulate the outcome of three-dimensional (3-D) multicellular morphogenesis. This phenomenon is particularly relevant for the design of materials suitable for supporting hepatocellular cultures, where cellular morphology is known to be intimately linked to the functional output of the cells. This review summarizes recent work describing biophysical regulation of hepatocellular morphogenesis and function and focuses on the manner by which biochemical cues can concomitantly augment this responsiveness. In particular, two distinct design parameters of the substrate biophysics are examined – microtopography and mechanical compliance. Substrate microtopography, introduced in the form of increasing pore size on collagen sponges and poly(glycolic acid) (PGLA) foams, was demonstrated to restrict the evolution of cellular morphogenesis to two dimensions (subcellular and cellular void sizes) or induce 3-D cellular assembly (supercellular void size). These patterns of morphogenesis were additionally governed by the biochemical nature of the substrate and were highly correlated to resultant levels of cell function. Substrate mechanical compliance, introduced via increased chemical crosslinking of the basement membrane, Matrigel, and polyacrylamide gel substrates, also was shown to be able to induce active two-dimensional (2-D, rigid substrates) or 3-D (malleable substrates) cellular reorganization. The extent of morphogenesis and the ensuing levels of cell function were highly dependent on the biochemical nature of the cellular microenvironment, including the presence of increasing extracellular matrix (ECM) ligand and growth-factor concentrations. Collectively, these studies highlight not only the ability of substrate biophysics to control hepatocellular morphogenesis but also the ability of biochemical cues to further enhance these effects. In particular, results of these studies reveal novel means by which hepatocellular morphogenesis and assembly can be rationally manipulated leading to the strategic control of the expression of liver-specific functions for hepatic tissue-engineering applications.

Keywords Hepatocellular · Morphogenesis · Substrate biophysics · Topography · Mechanical compliance

Abbreviations

GF	Growth Factor
EGF	Epidermal Growth Factor
HGF	Hepatocyte Growth Factor
FN	Fibronectin
PBS	Phosphate-buffered Saline
DMEM	Dulbecco's Modified Eagle Medium
BISAAm	methylenebisacrylamide
AAm	acrylamide

1
Introduction

The biophysical properties of the cellular microenvironment have been recognized to be critical in determining the outcome of cell morphogenesis and resultant expression levels of cell function [1-4]. Here, sensory mechanisms of cells are believed to be able to feel the physical structure of their proximal surroundings, causing cells to adopt a specific phenotype in response, which, in turn, may morphologically prime ensuing cell behavior [5-7]. While, in recent years, several useful correlations of this nature have been developed for cultured hepatocytes in the scope of liver tissue-engineering applications [8-14], the ability to rationally manipulate biophysical regulatory mechanisms by utilizing additional biochemical cues in the design of hepatocellular constructs has yet to be thoroughly explored. To this end, this review aims to present recent evidence that differential cell responsiveness to similar biophysical states can be achieved, particularly highlighting how the effects of two major design parameters of the substrate biophysics – microtopography and mechanical compliance – on hepatocellular behavior can be considerably altered by varying degrees of substrate adhesivity as well as by soluble stimulatory factors. Insights gained from these strategies may be useful, not only for developing highly efficient bioartificial liver devices and clinically tunable scaffolds for partial cell transplantation therapies [15-17], but also potentially for achieving effective stem-cell differentiation into mature, functional hepatocytes [18].

The importance of suitable 3-D architectures for hepatospecific support scaffolds has been documented in several recent studies [19-22]. In particular, the pore size and structure of the substrate have been demonstrated to be intimately related to the scaffold performance, providing not only a means of nutrient and waste transport, but also a surface containing microscale topography on which the cells reside, providing biophysical cues otherwise known as *contact guidance* for subsequent cell behavior. While generalized trends regarding pore size and hepatocellular response have been well established, the effects elicited by these cues can be diverse based on the scaffold chemistry and the degree of adhesivity or type of intracellular signal activation simultaneously induced by the cell-bound surface. For example, Moghe and coworkers systematically evaluated hepatocyte morphogenetic and functional behavior following the progressive incorporation of microtopography, in the form of increasing void size, on ligand-deficient poly(lactide-co-glycolide) (PLGA) polymer foams [23]. On such substrates, the topographic regimen (supercellular-scale pore size) that yielded the formation of multicellular spheroidal aggregates, maximizing 3-D cell–cell interactions, was reported to promote liver-specific functional activity. In contrast, in subsequent work, Moghe and coworkers found that varying microtopography on highly adhesive, ligand-presenting matrices such as collagen foams resulted in a markedly different trend in hepatocyte response [24]. Here, the role of cell–substrate interactions appeared to assume greater importance in functional regulation, as intermediate-scale topography (cellular-scale pore size) was demonstrated to *lower* the resultant functional activity by promoting cell spreading. These results are summarized schematically in Fig. 1.

Another substrate biophysical property that has received increasing attention in the design of hepatocellular constructs is substrate rigidity. Since cells are believed to be mechanosensitive, physical resistivity of the cell surroundings may govern cell shape and the output of future signaling processes. Here, there is an intrinsic relationship between substrate mechanical compliance and substrate adhesivity, since cells primarily perceive the substrate through anchorage points established via receptor–ligand complexes, through which cell tractile forces act being both applied and resisted. Studies in Moghe's laboratory recently demonstrated that increasing gel rigidity typically causes hepatocellular spreading on ligand-presenting polyacrylamide hydrogels, although this effect is clearly related to the ligand chemistry of the substrate [25]. Here, the presence of decreasing concentrations of the highly adhesive extracellular matrix (ECM) ligand, fibronectin, was demonstrated to elevate the functional output of hepatocytes cultured on the surfaces of differentially compliant gels. These effects were most pronounced for regimens of intermediate compliance and low ligand density. These findings are summarized schematically in Fig. 2.

Finally, using hepatotrophic growth factors, the morphogenetic and functional responses of hepatocytes to substrate compliance were significantly intensified by the presence of hepatocyte growth factor (HGF) and epidermal growth factor (EGF) [26]. In particular, cultures devoid of growth factor (GF) stimulation, but exposed to strikingly different levels of substrate rigidity, ex-

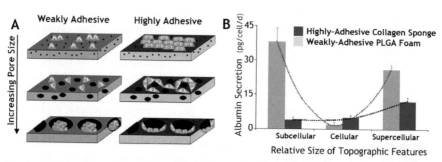

Fig. 1 **A** Differential patterns of hepatocellular morphogenesis induced by microporous substrates of varying substrate adhesivity. Pore size varied from subcellular-scale to cellular-scale to supercellular-scale, and differential substrate adhesivity was established by using substrates of unlike chemistry (weakly adhesive – PLGA foam; highly adhesive – collagen sponge). **B** The effects of substrate microtopography on the expression of hepatocellular function are dependent on the adhesivity of the substrate

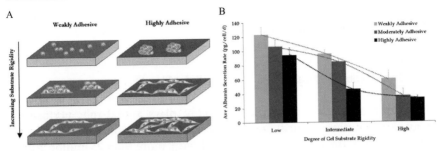

Fig. 2 A Diverse patterns of hepatocellular morphogenesis induced by differentially compliant substrates of varying substrate adhesivity. Substrate rigidity was altered via increased crosslinking of polyacrylamide gels during polymerization over a wide range, and differential substrate adhesivity was established by immobilizing increasing concentrations of the adhesive, proliferation-inducing ECM ligand, fibronectin, to the gel surface. **B** The nature of substrate adhesivity, mediated by varying concentrations of the ECM ligand fibronectin (FN), can shift the curve describing the effects of substrate rigidity on the expression of hepatocellular function

hibited minimal morphologic and functional differences between them, while cultures stimulated with both GFs exhibited drastically divergent responses in both morphogenesis and the expression of liver-specific activity. These results are summarized schematically in Fig. 3.

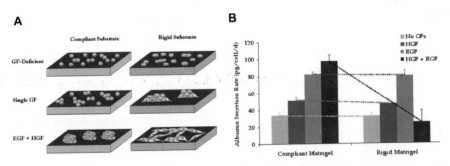

Fig. 3 A Distinct morphogenetic responses of hepatocytes cultured on compliant and rigid Matrigel in the presence of varying degrees of soluble growth factor (GF) stimulation. Notably, the most dramatic differences in hepatocellular morphogenesis between cells on compliant and rigid Matrigel occurred under GF co-stimulatory conditions. **B** Between compliant and rigid Matrigel, the effects of GF activation on the expression of hepatocellular function become divergent only under co-stimulatory conditions

2
Background and Review

2.1
Clinical Motivation for Engineering Hepatocellular Tissues

The liver, the largest organ inside the body, plays a vital role, performing many essential physiological functions such as metabolism, excretion, secretion, detoxification, and storage [27]. Each year more than 25 million Americans are afflicted with liver and gallbladder diseases and more than 43 000 fatalities and associated costs exceeding $16 billion are recorded annually resulting from liver failure [28]. Although the liver does possess some inherent regenerative capabilities, eventually self-restoring a degree of liver functionality, extensive tissue damage often compromises the natural recuperative abilities of the liver and prevents successful recovery from liver injury or disease [29, 30]. While whole-organ transplantation remains the only viable long-term solution to liver dysfunction currently available and the number of liver transplants preformed annually has increased nearly tenfold over the past decade, a severe shortage of available donor organs along with spiraling costs and high risks associated with surgery limit the widespread effectiveness of this treatment option [31, 32]. In addition, many biocompatibility issues between donors and recipients also exist which warrant consideration, and even after a successful transplantation, the five-year survival rate is only 65% [33, 34]. In order to develop alternative strategies, hepatic tissue engineering seeks to create structurally and metabolically functional cell-based substitutes for liver dysfunction, thereby overcoming shortcomings associated with traditional tissue-replacement therapies [35, 36]. This process is particularly challenging since the diverse physiological functions and high metabolic activity of the liver represent significant challenges to engineering devices that provide stable and effective hepatic support [37, 38]. Typically, in order to overcome these hurdles, liver tissue-engineering strategies incorporate the usage of hepatocytes, the liver parenchymal cells that perform most liver functions and make up 80% of the liver cytoplasmic mass as a cell source [39]. In recent years, much attention has also been directed to sourcing and utilizing differentiated stem cells for potential hepatocellular therapies [40–42]. Promising tissue-engineered applications currently under investigation include extracorporeal bioartificial liver (BAL) devices that utilize hepatic cells to provide temporary support for patients in acute liver failure and implantable systems that use transplanted hepatocytes inside biocompatible scaffolds that are designed to provide structural and functional support, creating tissue analogs once inserted into the body [43–46]. As hepatocytes are critically governed by their local substrate, one of the major research thrusts in the field focuses on controlling the properties of the hepatocyte-anchoring substrate, so as to design and control the organization and differentiated functional expression of the cells in the extracorporeal device, or upon implantation.

2.2
Matrix Biophysics as a Determinant
of Hepatocellular Organization and Function

The ultimate success of designing hepatospecific construct microenvironments relies on the ability to establish and sustain multicellular networks, so that cellular growth and expression of liver-specific functions can be attuned to the metabolic needs of the relevant liver pathology [47]. To this end, studies in hepatic tissue engineering have yielded three major classes of environmental regulatory variables that may play vital roles in designing successful hepatospecific constructs: soluble factors that may include nutrients, hormones, and growth factors; homotypic and heterotypic cell–cell interactions; and the acellular microenvironment to which cells are adhered such as the extracellular matrix (ECM) or substratum [48–51]. This microenvironment may be of particular importance since its biophysical makeup as well as its biochemical nature may regulate cellular attachment and morphogenesis as well as transmit regulatory signals to cells [52]. The mechanical rigidity of the surrounding matrix, as well as material roughness and physical confinement, determined by three-dimensional microstructure on a subcellular- and supercellular-scale, respectively, may significantly modulate the outcome of the balance between cell–matrix forces, leading to the remodeling of cytoarchitecture, cell polarization, alteration of downstream intracellular signaling events as well as modification of the balance of cell–cell forces [12, 53, 54]. This chain of events is mediated by the cellular cytoskeleton (CSK), which is hardwired to its surrounding environment through a multitude of transmembrane receptors that typically bind specifically to various ECM molecules and neighboring cells [55, 56]. Therefore, as cells recognize and respond to external mechanical forces, they are forced to remodel their CSK until the local stress is minimized, resulting in cellular morphogenesis [57]. For example, cells on a rigid substrate versus a compliant substrate will change shape in a dramatic manner due to differences in physical resistivity (see Fig. 4).

Dynamic changes in the architecture of the CSK can result in significant alterations of the intracellular signaling cascades that govern cell growth, differentiation, motility, and gene expression [58]. Here, it is postulated that transmembrane receptors, which physically couple internal CSK networks to external support scaffolds or neighboring cells, are linked to specific molecular pathways for mechanical signal transfer across the cell surface in units such as the focal adhesion complex (FAC) [59]. Inside the FAC, a variety of intracellular signaling molecules are physically immobilized (ion channels, protein kinases, and lipid kinases) or lie in the main path of mechanical

A Matrix Reorganization   **B Cell Elongation**

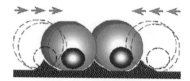

Fig. 4 Variations in the mechanisms of biophysically regulated cellular morphogenesis: active matrix reorganization on a substrate that is mechanically compliant to cell tractile forces (**A**) versus cell elongation on a substrate that provides sufficient physical resistivity to cell tractile forces (**B**)

force transfer, leading to the translation of mechanical stresses into biochemical responses [60, 61]. Additionally, the FAC also represents a potential site for signal integration between mechanosensitive responses and those arising from the activation of specific cell receptor–ligand transduction cascades, including those originating from soluble GF stimulation (receptor tyrosine kinases) as well as components of the ECM (integrins), prior to effecting the output of the downstream metabolic machinery of the cell [62, 63].

2.3
Biophysical Environment of Hepatocytes in Vivo

To gain insight into the biophysical regulation of hepatocytes, it is important to consider the biophysical nature of the surrounding hepatocellular environment in vivo. This may be particularly relevant since one potential difficulty in artificially reproducing physiological hepatocellular activity in vitro is that, under classical culture conditions, the cells are exposed to a 2-D matrix environment. This culture configuration typically induces a cell phenotype that may restrict normal regulatory processes as opposed to that observed in native 3-D liver structure. Here, hepatocytes are arranged in hepatic plates, which are separated by large capillary spaces, called sinusoids, inside liver lobules (Fig. 5). In this configuration, hepatocytes are highly polarized cells that are approximately 30 μm in diameter and polygonal in shape, contacting neighboring cells on two surfaces while being flanked by ECM [64]. Further evidence in vivo supporting the hypothesis is noted during liver regeneration after partial hepatectomy. Here, it is observed that the onset of liver injury induces hepatocytes to become physically loosened from each other and their surrounding matrix, a process that enables cellular proliferation at the expense of expression of liver-specific function [65, 66]. Similar trends in cell response have also been reported during episodes of increased collagenase expression in the liver, further suggesting that native biophysical signals as well as biochemical ones may be required to sustain physiological levels of liver function [67].

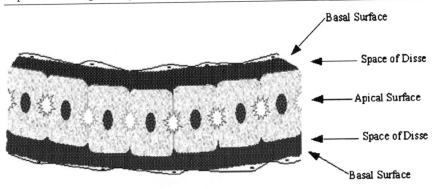

Fig. 5 Arrangement of hepatocytes in native liver [143]. Hepatic cells are enclosed in plates, one cell in thickness flanked on both surfaces by the extracellular matrix inside the space of Disse and normally 6–12 cells in number. This arrangement yields distinct apical (bile canalicular) and basal (sinusoidal) surfaces that serve different functions unlike classical epithelium. Therefore, current trends in hepatic tissue engineering include the development of techniques to modify the cellular microenvironment in vitro in order to achieve more favorable configurations of hepatocytes that may lead to increased levels of differentiated activity [11]

2.4
Early Evidence for Biophysical Sensitivity
of Hepatocellular Behavior in Vitro

Early examples of the importance of biophysically controlled cell shape were apparent in conventional cell-culture substrates for hepatocytes that typically consisted of ECM proteins such as collagen and basement membrane complexes [68, 69]. For example, when freshly isolated hepatocytes were plated on rat tail collagen, cells exhibited a flat, extended morphology, actively synthesized DNA and expressed high levels of cytoskeletal mRNAs and proteins such as actin and tubulin, while expressing reduced levels of liver-specific mRNAs and limited secretion of albumin [12, 70]. However, if another layer of collagen was gelled above the cells to change the geometry of the microenvironment, known as a collagen sandwich culture, hepatocyte morphology became more cuboidal in nature, cells were repolarized, and higher levels of liver function were recovered [71, 72]. Another example of biophysical modification of the collagen culture system was reported in studies where collagen was denatured, producing a malleable substratum as a cellular interface [73]. On these compliant surfaces, hepatocytes adopted a more spheroidal morphology and exhibit increased levels of metabolic activity versus cultures on native collagen. The notion that substrate rigidity may strongly influence cell behavior may also partially explain the effectiveness of another popular hepatocellular culture system, commonly known as Matrigel. Matrigel, a highly malleable hydrated gel reconstituted from Engelbreth–Holm Swarm (EHS) mouse tumor, induces spheroidal aggregate formation and causes cells to undergo low DNA, cytoskeletal mRNA, and protein synthesis, while at the same time exhibiting elevated liver-specific mRNAs and albumin production similar to native levels [74, 75]. This premise was further supported by studies demonstrating that chemical crosslinking of Matrigel reduces its capacity to establish a highly functional hepatocellular phenotype [26, 76].

2.5
Hepatocellular Morphology Intimately Linked to Functional Expression

In recent years, cell morphology has been strongly correlated to cellular behavior for a wide variety of cell types [77, 78]. In many cases, specific cell phenotypes have been linked to distinct patterns of cell survival, proliferation, motility, differentiation, and functional gene expression. This paradigm has been commonly observed for hepatocytes in vitro [9, 10, 13, 14, 79, 80], particularly for culture systems that have diverse biophysical attributes. Here, as previously described, patterns of hepatocellular morphogenesis evolve in a manner that appears to be highly correlated to differentiated function. Notably, hepatocyte cultures, which are three-dimensional in nature exhibiting a compacted, spheroidal morphology, typically express elevated levels of liver-specific functions indicative of a highly differentiated state when compared to cultures that have adopted a more two-dimensional morphology [11, 81]. Here, spread or elongated cells maintain a phenotypic state typically associated with increased cell growth behavior [80]. These observations reinforce a reciprocal relationship between hepatocyte growth and

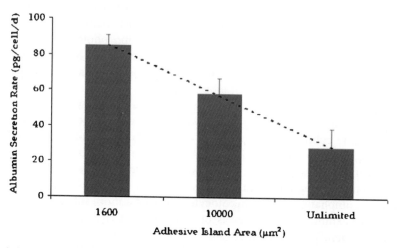

Fig. 6 Increased cellular spreading reduces the expression of liver-specific function [13]. Single hepatocytes were cultured on adhesive islands of varying dimensions, thereby restricting the area over which cells could spread

differentiation: conditions under which cells express high levels of hepatospecific function elicit low tendencies for proliferation and vice versa [82, 83]. The first direct evidence of the relationship between hepatocyte shape and function was reported by Singhvi et al., who demonstrated that the degree of spreading and differentiated function of hepatocytes plated on variably sized adhesive islands were highly (inversely) correlated (Fig. 6) [13]. Parallel studies that varied design parameters other than those that are biophysical in nature (such as ligand presentation [8, 84], surface hydrophobicity [9], and surface wettability [85]) have further reinforced the hypothesis that hepatocellular morphology can determine cell functional fate. More recently, the cellular phenomena that may underlie shape-induced regulation of hepatocyte function have been more rigorously probed by examining correlations between cellular activity and membrane dips over nuclei [86] as well as cytoskeletal stiffness for differentially spread cells [87].

2.6
Varied Approaches for Biophysical Manipulation of Hepatocellular Substrates

Over the past decade, researchers have explored the effects of diverse biophysical properties of the culture microenvironment on hepatocellular behavior with the ultimate goal of establishing stable cultures of hepatocytes capable of exhibiting sustained, physiological levels of liver-specific function [11, 88]. These systems are often designed to induce changes in the spatial arrangement of hepatocytes in vitro by varying a wide range of biophysical features such as geometry, topography, spatial confinement, and mechanical rigidity. A summary of recent studies reporting altered hepatocellular morphogenesis and function upon modification of substrate biophysics is tabulated below (Table 1). Notably, culture substrate geometry has typically been modified by the introduction of additional cell-interactive surfaces such as *sandwiching* cultures or culturing cells within 3-D structures rather than 2-D films. Substrate topography has been varied by augmenting substrate pore size from that below the cellular scale to a supercellular-scale level. Spatial confinement of the cells is often introduced by adjusting mesh size in fabric-like materials or by completely confining cells via encapsulation. Finally, substrate mechanical compliance is commonly varied by changing the degree of crosslinking in hydrogel or biopolymeric substrates. Collectively, these studies demonstrate that a wide range of biophysical properties, relevant to the design of hepatospecific constructs for liver tissue-engineering applications, can significantly modulate hepatocyte morphogenesis and provide further evidence of morphological control of liver-specific function.

Despite the studies tabulated above, few efforts have systematically investigated the regulation of cell behavior via substrate biophysical properties as a function of substrate biochemistry. For the remainder of this review, we

Table 1 Summary of recent literature reporting variations in hepatocellular morphogenesis and function via biophysical manipulation of the cellular microenvironment

Biophysical property studied	Substrate material(s)	Method of variation	Induced morphology	Functional expression	Refs.
1-d substrate geometry	Collagen gel	Replating	Spheroidal disassembly	Decreased	[89]
2-D substrate geometry	Collagen gel	Sandwich	Cuboidal	Increased	[12, 79, 90, 91]
2-D substrate geometry	Collagen gel	Matrigel overlay	Cuboidal	Increased	[79]
3-D substrate geometry	Alginate	Sponge vs. film	Spheroidal	Increased	[20]
3-D substrate geometry	Polyurethane	Foam vs. film	Spheroidal	Increased	[92]
Substrate topography	Collagen foam	Increasing pore size	Elongated then semi-spheroidal	Biphasic	[24]
Substrate topography	PLGA	Increasing pore size	Spheroidal	Increased	[23]
1-D spatial confinement	Laminin adsorbed to hexadecanethiol	Increasing adhesive island size	Spread	Decreased	[13]
2-D spatial confinement	PVLA-coated PET membranes	Increasing mesh size	Spheroidal	Increased	[21, 93]
2-D spatial confinement	Poly(hyaluronic acid ester) fabric	Increasing mesh size	Spheroidal	Increased	[94]
3-D spatial confinement	Collagen gel	Entrapment	Spheroidal	Increased	[95]
3-D spatial confinement	Agarose hydrogel	Encapsulation	Spheroidal	Increased	[96]
3-D spatial confinement	Collagen-coated acrylic composite	Encapsulation	Spheroidal	Increased	[97]

Table 1 (continued)

Biophysical property studied	Substrate material(s)	Method of variation	Induced morphology	Functional expression	Ref.
Substrate flexibility	Matrigel	Increasing gel thickness	Spheroidal	Increased	[98]
Substrate rigidity	Collagen	Denaturation	Spheroidal	Increased	[73]
Substrate rigidity	Matrigel	Chemical crosslinker after gelation	2-D Elongated	Dependent on GFs	[26]
Substrate rigidity	FN-presenting polyacrylamide gels	Increasing crosslinker Concentration upon polymerization	2-D elongated	Fibronectin-dependent decreased	[144]
Ligand micromobility	PEO-tethered galactose	Increasing tether length	Spreading	Not reported	[99]

will focus on two case studies describing the potent, yet diverse, ability of two separate biophysical design parameters, substrate microtopography and substrate mechanical compliance, to govern hepatocellular morphogenesis and the resultant expression of liver-specific function. The roles of substrate physical properties on cell responses will be examined for different levels of substrate biochemistry (mediated through diverse ligand chemistry and density) as well as soluble stimulation (mediated through the growth factors EGF and HGF). Findings reported in these studies not only highlight biophysical regulation of hepatocellular behavior but also demonstrate the ability of biochemical factors to either enhance or repress these effects.

3
Case Study #1 –
Role of Substrate Microtopography in Hepatocellular Engineering

3.1
Design

3.1.1
Selection of Microporous Biocompatible Scaffolds for Hepatocyte Culture

In order to examine hepatocyte morphogenetic and functional response to varying surface microtopography, two substrate systems with distinct bioadhesive surfaces were evaluated in which scaffold surface features could be readily manipulated during fabrication. The first, based on porous collagen sponges, was adapted from one of the early three-dimensional scaffold-type configurations developed from ECM proteins, being originally intended to be membranes for full-thickness wound closure, promoting cellular infiltration, vascularization, and epithelialization [100–102]. While these porous collagen matrices have widely been investigated for their ability to support various tissues [103–107], there is limited understanding of the key regulatory parameters of hepatocyte function in a highly adhesive, microporous collagen environment, particularly with regard to microstructural geometry.

The second culture system consisted of synthetic polymer matrices such as the poly(α-hydroxy ester) family of polymers, which represent a prime example of substrates that exhibit generic biocompatibility yet, unlike collagen sponges, may lack biochemical binding moieties to promote hepatocellular differentiation. Copolymers of poly(lactic acid) and poly(glycolic acid) (PGLA) from this family have been previously employed for hepatocyte support, both in vitro and in vivo [16, 108–110]. While various structural PGLA templates have been used for hepatic tissue engineering, the role of matrix topography and topology in eliciting controlled variations in hepatocyte organization and functional activities has not been systematically elucidated.

Given their lack of hepatospecific chemistry, these PGLA substrates can serve as a model for microstructural versus biochemical regulation of hepatocellular differentiation.

3.1.2
Design of Biomaterial Scaffolds with Variable Surface Microstructure

For the first culture system, collagen type I matrices were prepared from bovine deep flexor tendon, dispersed at 1–3% w/v in alkali solution (Integra LifeSciences Corp., Plainsboro, NJ). Following de-aeration, the collagen solutions were subjected to various freezing rates to produce foams with average pore sizes ranging from $< 10\,\mu m$ to $100\,\mu m$ in diameter. Frozen structures were freeze-dried to sublimate the frozen liquid phase, leaving behind porous collagen-based scaffolds, which were subsequently stabilized via formaldehyde crosslinking or ultraviolet (UV) irradiation and sterilized using ethylene oxide. Collagen matrices were fabricated as sterile 1–2-mm-thick sheets of varying dimensions (Integra LifeSciences Corp., Plainsboro, NJ). Circular disks were prepared from this material under sterile conditions using 6-mm Biopsy punches (Fray Products Corp., Buffalo, NY), immobilized in 96-well plates, and soaked in complete culture medium for 2–4 hours prior to further study. The pore morphology of these scaffolds was assessed with laser scanning confocal microscopy. Reflected-light images of the surface of each of the collagen foams were digitally acquired, and subsequently analyzed using Image Pro software to determine the average pore diameter and pore-size distribution of each sample. Specifically, three void size configurations were designed to induce pronounced changes in hepatocyte functions via a unique progression of cell morphogenesis. The distinct pore-size configurations of the three primary classes of collagen foam were visually confirmed using reflective-mode confocal laser scanning microscopy. Values for the average pore diameter were determined to be $9.9 \pm 0.47\,\mu m$, $17.7 \pm 3.2\,\mu m$, and $81.6 \pm 3.5\,\mu m$, respectively, for the collagen foams represented in Figs. 7a, b, and c, respectively. The percent porosity of each of these three collagen foam configurations was estimated as 90–94%.

For the second culture system, three variably textured polymer foams were prepared from 50/50 poly(DL-glycolic-co-lactic acid) copolymer (PGLA) using a phase-separation technique [111]. The weight-averaged molecular weight (Mw) and polydispersity index (Mw/Mn) of the PGLA was specified as 7.5×10^4 and 1.95, respectively (MEDISORB®, Alkermes). Resulting PGLA foams were primarily analyzed for differences in surface topography. Topographical measurements were performed using reflected imaging on a confocal laser scanning microscope (LSM). The morphology of the surface pores of PGLA-based foams was quantitatively characterized in a similar manner to that of collagen sponges as earlier described. Topography profiles were generated using the topography scan function in the LSM software, which cal-

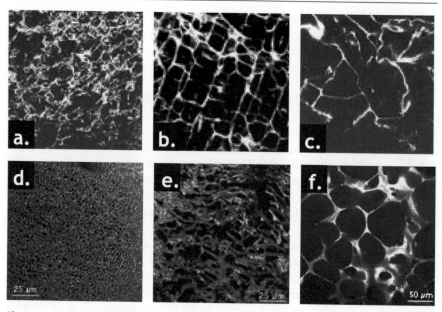

Fig. 7 Pore morphology of collagen type I-based sponges (**a–c**) and PLGA foams (**d–f**) generated using reflection confocal laser-scanning microscopy: collagen sponges with average pore diameters of **a** 9.9 μm, **b** 17.7 μm, **c** 81.6 μm; and PLGA foams with average pore sizes of **d** 2.7 μm; **e** 16.5 μm; and **f** 67.4 μm

culates the height value corresponding to each pixel intensity relative to the maximum intensity. Typically, samples were optically sectioned into 10 sections, each 1 μm apart along the z-axis. Topography profiles were smoothed with the use of median and low-pass filters to generate the final images. Optical sections were further analyzed to quantitatively evaluate the degree of surface roughness by applying the principle that the maximal intensity of reflected light corresponds to an object exactly in the focal plane [112]. Thus, intensity values at each x, y coordinate were compared at various depths to derive local height values, z_i, which were used to compute the mean height, z_m, and finally, the root-mean-squared roughness value, R_q:

$$z_m = \frac{\sum_i z_i}{N}, \quad R_q = \sqrt{\left\{ \left[(z_1 - z_m)^2 + (z_2 - z_m)^2 + ... \right] / N \right\}},$$

where z_i is the height value corresponding to x_i, y_i, and N is the number of pixels in a single plane. Confocal reflection micrographs of the three PGLA foam configurations showed considerable differences in the size of surface pores (Fig. 7d–f). Quantitative analysis of the three resultant polymer surfaces revealed a uniform surface structure with good control over the pore-size distribution (data not shown). Pore-size distributions computed for the three conditions had average pore sizes, ϕ, of 2.7 ± 0.9 μm, 16.5 ± 6.6 μm, and

67.4 ± 23.7 μm, respectively. The quantitative degree of surface roughness, R_q, was computed from topography profiles to be: 1.03 ± 0.03, 1.90 ± 0.02, and 2.44 ± 0.01 μm, respectively, for the 2.7, 16.5, and 67.4 μm foams, while R_q values for the untextured thin films (control) were 0.018 ± 0.001 (scans not shown).

3.2
Results

3.2.1
Evaluation of Cellular Morphogenesis and Multicellular Assembly

For hepatocytes cultured on collagen foams, the variations in pore structure significantly modulated cellular organization on the collagen foams after initial adhesion, although the degree of cell attachment to collagen matrices was not affected by pore structure. In order to probe cellular morphogenesis, cell shape was assessed in terms of the cytoskeletal organization by fluorescently labeling of F-actin with fluorescein phalloidin (Molecular Probes, OR), which binds to cytosolic F-actin microfilaments. Hepatocytes seeded onto 10-μm collagen foams were physically excluded from the interior of the scaffold, resulting in the formation of discontinuous trabecular arrays along the surface. During the first week of culture, hepatocytes actively reorganized, ultimately achieving monolayered surface coverage (Fig. 8a). Individual hepatocytes comprising these monolayers were found to maintain a cuboidal, compact morphology throughout the culture period. Hepatocytes cultured on 18-μm foams were also observed to attach predominantly along the surface of the porous foams rather than throughout the network of small pores. During the first week of culture, hepatocytes established discontinuous sheets of irregularly shaped cells along the foam's surface (Fig. 8b). In

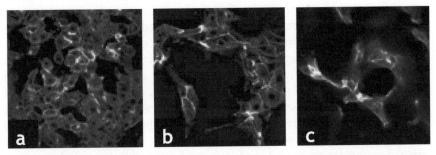

Fig. 8 Hepatocyte morphogenesis on **a** 9.9-μm, **b** 17.7-μm, and **c** 81.6-μm collagen sponges after three days of culture. The cytoskeletal protein F-actin was fluorescently labeled with fluorescein phalloidin, and cells were fluorescently viewed at 40× magnification using a laser scanning confocal microscope

contrast, hepatocytes attached to the 80-μm foams as small, 3-D aggregates along and within the pores. During the first week of culture, these cells began to align circumferentially along the edges of the pores, infiltrating to greater depths, ultimately establishing a complex banded 3-D cellular network (Fig. 8c). The cells comprising this network exhibited an extremely spread phenotype.

Hepatocyte morphology was quantitatively assessed via digitized image acquisition and analysis via Image Pro Plus software (Media Cybernetics, Silver Spring, MD). In this manner, hepatocytes cultured on 10-μm collagen sponges were determined to exhibit the least 2-D spreading, whereas those on 18-μm sponges were measured to have the greatest area of spreading (Fig. 10a). Moreover, the average area of cells cultured on the 10-μm sponges increased temporally at the rate of 60% at day 3, and subsequently decreased to initial values by day 5. This biphasic behavior was well supported by our visual observations, as earlier described. Here, hepatocytes initially attached to the foams in a compact, spherical morphology, and subsequently adopted a more spread phenotype, as cell–cell contacts were actively established. Once a confluent monolayer was achieved, hepatocytes compacted considerably, and appeared cuboidal in shape. Hepatocytes cultured on 80-μm sponges adopted a significantly more spread phenotype (55%) in comparison with 10-μm foam cultures. These cells continued to spread (22%) during the next two days, and maintained this spread morphology for the duration of the culture. Similar behavior was observed for cells on 18-μm sponges, as the average cell area increased 42% between days 1 and 3, with no further changes at later time points.

Hepatocyte multicellular organization on PGLA foams varied significantly with time, and in response to the underlying polymer pore structure. Over the first two hours post seeding, hepatocytes on 2.7-μm foams actively reorganized along the surface to form large clusters of weakly associated cells (not shown). In contrast, hepatocytes seeded on the 16.5-μm and 67.4-μm foams exhibited a lower extent of multicellular reorganization during this time interval, and showed a significantly lower degree of aggregation (not shown). These differences in topography-mediated hepatocyte morphogenesis became more pronounced at later time intervals. After 72 h of culture, the largest hepatocyte aggregates (in terms of surface area) were observed on the 2.7-μm foams (Fig. 9a–c). Here, the extent of hepatocyte aggregation progressively decreased with increasing pore size, and the differences in the aggregate morphology on the three PGLA foams evolved to be quite striking.

For PGLA scaffolds, hepatocyte reorganization behavior was quantitatively determined by measuring aggregate morphology at specific time intervals. Here, hepatocytes on the 2.7- and 16.5-μm foams exhibited a slight increase in the average aggregate area over time, whereas aggregates on 67 μm foams showed a decrease in area by day 3 (Fig. 10b).

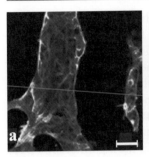
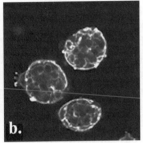
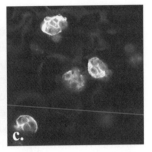

Fig. 9 PGLA microstructure-induced hepatocyte morphogenesis after three days of culture: fluorescent confocal images (40× magnification) of F-actin distribution in hepatocytes cultured on PGLA foams with **a** 2.7-μm, **b** 16.5-μm; and **c** 67.4-μm pores. The *scale bar* represents 50 μm

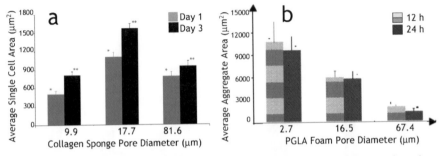

Fig. 10 a Cell spreading based on the average single cell area for hepatocytes cultured on the 9.9-μm, the 17.7-μm; and the 81.6-μm collagen sponges. *Single asterisks* indicate statistically significant differences between day 1 samples ($p < 0.05$) while *double asterisks* indicate statistical significance among day 3 samples. **b** Aggregate area computed 12 and 24 h post seeding for hepatocytes cultured on the 2.7-μm, 16.5-μm, and 67.4-μm PGLA foams. The *asterisk above each bar* indicates statistically significant differences between measurements at the given time point. For all measurements, *error bars* represent the standard error around the mean

3.2.2
Expression of Liver-Specific Function

For both culture systems, the metabolic activity of hepatocyte cultures was evaluated by measuring the levels of albumin secreted into the culture medium. The albumin concentration of the daily culture samples was determined using an enzyme-linked immunosorbant assay (ELISA), using purified rat albumin standards and peroxidase-conjugated anti-rat antibodies (ICN Biomedicals, OH). Albumin secretion rates were normalized to the total number of cells attached to each culture substrate, or to the DNA content of each sample, which was measured by using Hoechst dye 33250 (Molecular Probes) [113]. For hepatocyte cultures on collagen foams, values of albumin secretory rates were significantly affected by the collagen pore-size configuration employed for hepatocyte culture (Fig. 11a). The albumin secretion rates of 18 μm cultures were consistently low, and closely paralleled those of the control cultures (single collagen gel). Significant increase in albumin secretory activity was observed with cultures on both 10-μm and 80-μm foams, where average albumin secretion rates reached 40 pg/cell/day for the 10-μm cultures versus 26 pg/cell/day for the 80-μm cultures during the second week of culture.

Early values of albumin secretion kinetics were not significantly different on the three PGLA foam configurations studied (Fig. 11b). However, a gradual, statistically significant, increase in average albumin secretion rate was observed later in the culture (days 4–8). Hepatocytes cultured on the 67.4-μm foams consistently exhibited significantly higher albumin secretory activity than those cultured on the 16.5- or 2.7-μm foams ($p < 0.05$), reaching

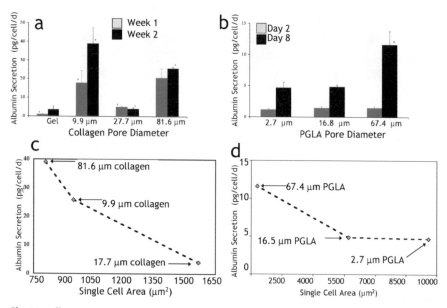

Fig. 11 Albumin secretion rates of hepatocyte cultures established on collagen (**a**) and PGLA (**b**) substrates of distinct microstructure at various time points. Collagen substrates included a single layer of collagen I gel, 9.9-μm collagen sponges, 17.7 μm collagen sponges, and 81.6-μm collagen sponges. PGLA substrates included 2.7-, 16.5-, and 67.4-μm PGLA foams. *Error bars* represent the standard error around the mean, and *asterisks* indicate statistically significant differences between samples ($p < 0.05$) at a given time point. Relationship between albumin secretion rates and cell morphology on collagen sponges (**c**) and PGLA foams (**d**) in terms of single cell area and average aggregate area, respectively

maximum values of 12.9 ± 2.7 pg/cell/day ($n = 3$; day 7). Also, albumin secretion rates were better sustained on the 67.4-μm foams, where only a gradual decline in functions was observed at the end of the two-week culture period (data not shown). It should be noted that the maximal albumin secretion rate we report (day 8; 67.4-μm foams) is four times greater than the values reported previously for hepatocytes cultured on collagen gels (3.9 ± 1.2 pg/cell/day), but is significantly lower than cultures on other organotypic 2-D matrices such as Matrigel (66.0 ± 17 pg/cell/day) [12].

3.3
Conclusions

3.3.1
Dimensions of Substrate Microtopography Differentially Govern Hepatocellular Morphogenesis

These studies highlight the role of substrate microtopography of both highly biochemically active ECM-based substrates as well as relatively nonphysiological synthetic polymer matrices on the induction of both hepatocellular morphogenesis and ensuing expression of liver-specific function. For both collagen sponge and PGLA scaffold culture systems, three distinct magnitudes of median pore size were developed and used to gauge hepatocellular morphogenetic and functional response to surface topographical features. These magnitudes of median pore size offer a wide range of topography, representing the subcellular (3–10 μm) to the cellular (16–18 μm) to the supercellular (67–80 μm) scale. Although the intrinsic cellular adhesivity for each set of substrates in either culture system did not vary between topographical conditions, the choice of pore size did strongly impact the extent and level (2-D vs. 3-D) of hepatocyte interactions with other hepatocytes and with the substrate, leading to remarkably different long-term morphologic endpoints. Here, the three pore sizes were found to selectively alter cellular coverage, spreading, and multicellular aggregation for each culture system.

Although the collagen sponge substrates were highly biochemically active and strongly adhesive, the variations in pore size greatly affected the outcome of cellular adhesion and ensuing morphogenesis yielding quite distinct multicellular morphologic states. For example, on subcellular-sized pores, cells were restricted to the surface of the collagen foam, engendering a 2-D morphology and aggregate configuration. Moreover, the underlying pore texture also limited cell spreading, and promoted cellular compaction. This topography-induced cellular morphogenesis is markedly different from that elicited by hydrated collagenous substrates, on which cells exhibit pronounced flattening and spreading [114]. For cells cultured on 10 μm foams, for example, the cytoskeletal protein F-actin was observed to be mainly pericanalicular and at cell-cell contacts (also reminiscent of the distribution seen in compacted, collagen-sandwiched hepatocytes), which is in striking contrast to the F-actin containing stress fibers reported on single collagen gel cultures [91]. The use of cellular-scale pores (18-µm foams) in these studies precluded much cellular infiltration, and the limited degree of cell–cell contacts established with proximally available cells, resulted in discontinuous, patchy surface coverage. Under such conditions, hepatocytes exhibited irregular, highly spread morphology, with very low levels of peripheral and pericanalicular localization of F-actin. Overall, it is hypothesized that at both subcellular and cellular scales (10 and 18 µm respectively), the surface pores may primarily pose as significant surface discontinuities, which may strongly regulate cellular phenotype. In contrast, foam surfaces with the largest pore size (80 µm) favored cellular infiltration into the pores, and small, multicellular aggregates were established. Additionally, the absence of any cellular-scale topography further promoted cellular aggregation, facilitating the formation of extensive 3-D cellular networks along the entire foam surface. Thus, the pore size of collagen matrices can directly regulate cellular spreading and the underlying cytoskeletal organization, which, in turn, can govern multicellular aggregation behavior.

For the PGLA scaffold system, several effects of pore-associated topographical variations on resultant hepatocellular morphogenesis were similarly identified. Morphologic results indicate that topographic changes in microporous PGLA substrates can sensitively regulate the extent of multicellular spreading in both 2-D and 3-D. Here, the spatially regular variations in topography posed by the presence of micropores are thought to be equivalent to surface texture on a macroscopic, multicellular, scale [115]. The degree of surface roughness, R_q, of these substrates, reflects the spatially averaged variations in texture across the surface; this index increased with the size of pores, and was highest for 67.4-µm foams. An inverse correlation between R_q and the hepatocyte aggregate spreading factor (defined as the extent of aggregate spreading relative to the textural feature size of the substrate) was determined. Thus, substrates with the highest R_q values elicited the lowest aggregate spreading factor, indicating that aggregates no longer spanned multiple topographical features, but instead, were trapped within them, leading to large 3-D aggregates approximately spheroidal in nature. Here, cells may have limited initial 2-D contacts, but in the absence of any tight membrane contacts with the substrate, active cellular reorganization could have been less restricted, allowing the cells to repeatedly layer up to form 3-D aggregates. On the other hand, substrates with the lowest R_q values promoted the 2-D aggregation across multiple topographical features and did not allow 3-D cellular reorganization to occur. Here, on the 2.7-µm pore foam, the tight cell–substrate contact established on these substrates may provide traction for 2-D cell elongation, while precluding extensive cellular multilayering in 3-D.

3.3.2
Topography- and Adhesion-Induced Morphogenesis
Regulates Hepatocellular Function

In the collagen sponge system, the metabolic culture analysis demonstrated that hepatocyte albumin secretion rates varied significantly in response to the morphologic states arising from the varying foam microstructure between the types of collagen sponges (Fig. 11c). For these studies, the highest albumin secretory activity observed on subcellular scale (10-μm) foams is a unique instance of highly elevated function induced by collagen topography, and these levels are significantly greater than those reported in any other monolayer collagen cultures [69, 79]. It is believed that the underlying enhancement in liver-specific gene expression may arise from pore-induced compaction of cell shape and the establishment of native cell polarity, and to a lesser extent, may be due to a high degree of cell–cell contact establishment. The aspect in which these highly functioning cultures differ from previously tested collagen sandwich cultures is the absence of any collagen overlay, which has also been reported to similarly achieve restricted cell spreading and enhanced liver-specific polarity and functions [12, 71, 91, 116, 117]. On the other hand, using foams with pores sized at 18 μm, slightly below the average size of an isolated hepatocyte, resulted in a complete loss of function, similar to that observed with the monolayer collagen gel cultures. These results suggest that the lack of differentiation may occur primarily due to increased cell spreading within the interpore regions, phenotypically indicative of a switch from differentiation to *growth* behavior. The poor differentiation may also be secondarily due to the low degree of intercellular contacts, a critical level of which may be requisite to elevated gene expression [69, 82, 118]. Interestingly, hepatocytes on 80-μm foams regained their functions significantly in spite of a somewhat spread cellular phenotype and extended cell–matrix contacts characteristic of undifferentiated cells. The extent of cell–cell versus cell–matrix interactions can regulate liver-specific gene expression, suggesting that the 3-D cell–cell contacts established in this system may have compensated for the effects of the undifferentiated cell phenotype [83, 114, 119].

Variations in the topography of microporous PGLA substrates were also demonstrated to have the ability to sensitively control patterns of hepatocellular albumin secretion. Here, the changes in the expression of liver-specific function can be reconciled given the evolution in aggregate morphogenesis on the different foam substrates (Fig. 11d). In a similar fashion to that observed for hepatocyte cultures on collagen sponges, the expression of hepatocellular function markedly decreased with increased cellular spreading. In addition, the degree of cell–cell interactions may also effect liver-specific gene expression, as the cells aggregate in 2-D and 3-D on the foam substrates. The 67.4-μm foam configuration maximized cell–cell contacts and minimized the less favorable cell–substrate contacts, resulting in a 3-D aggregate morphology. The 16.7-μm foams supported both 2-D and 3-D reorganization, however, the 3-D thickness of the aggregates was lower that seen on the 67.4-μm foams. It is possible that the proximal cell–substrate contacts with the nonphysiological polymer may have limited the rise in albumin secretion kinetics seen on the 16.7-μm foams. Thus, overall, the observed increase in albumin secretion rates point to the importance of 3-D cell aggregation in promoting cell-to-cell contact-related signaling and *shielding* of the cells in the interior from the relatively nonphysiological polymer contacts. This is consistent with previous studies that have shown 3-D cellular organization to be essential, within limits, for improved hepatospecific functions [83, 120, 121], particularly when cells are cultured on substrates that lack hepatospecific chemistry [10, 88, 119, 122].

While topographically driven morphogenesis was demonstrated to either critically enhance or limit the degree of expression of liver-specific function for both culture systems, the manner by which this phenomenon occurred is itself highly dependent upon the biochemical nature of the underlying substrate. For a highly adhesive, bioactive substrate such as collagen, strategic modifications of pore architecture were demonstrated to be able to elicit the organotypic progression of hepatocytes, from a 2-D differentiated culture using 2-D and 3-D restrictive geometry (subcellular-scale micropores), to a 2-D undifferentiated culture using 2-D unrestrictive, but 3-D restrictive geometry (cellular-scale micropores), and finally to a moderately differentiated 3-D system using 2-D and 3-D unrestrictive geometry (supercellular-scale pores). Strategic manipulation of the voidage of PGLA foams significantly affected hepatocellular spatial organization in a somewhat different manner, resulting only in increasing aggregate spreading as void size decreased. This was in contrast to spreading behavior on collagen sponges where intermediate cell spreading was observed on the smallest-pore-size substrates. Although the response of cellular morphogenesis on PGLA substrates to void structure diverged from that on collagen sponges, the ensuing expression levels of differentiated function of surface-localized cultured hepatocytes had a strikingly similar dependency on the degree of hepatocellular morphogenesis. Here, cultures exhibiting the highest degree of cellular spreading expressed the least amount of liver-specific function, while cultures exhibiting the most compact morphology expressed the most. Unlike collagen sponge cultures, hepatocytes cultured on the PGLA system only exhibited increased levels of liver-specific function when exposed to the supercellular-scale-pore substrates. This configuration allowed substantial 3-D aggregate morphogenesis, not only affecting cellular shape but also perhaps increasing the number and nature of intercellular contacts in a manner favorable to the elevation of cell functional activity. Thus, in the absence of adhesive ligands, 3-D cell–cell contacts may be significantly more essential to the expression of liver-specific function than under conditions of sufficient substrate adhesivity, where semi-spheroidal or cuboidal cells may also give rise to high levels of cell differentiation.

4
Case Study #2 –
Role of Substrate Mechanical Compliance in Hepatocellular Engineering

4.1
Design

4.1.1
Selection of Differentially Compliant Hydrogel Substrates for Hepatocyte Culture

In order to examine hepatocyte morphogenetic and functional response to varying substrate mechanical compliance, two separate culture systems were developed in which the bulk rigidity of the underlying substrate could be readily manipulated during fabrication. The first culture system utilized for this investigation was based upon the highly adhesive basement membrane hydrogel, Matrigel, while the second substrate system comprised basally non-adhesive polyacrylamide gels that were functionalized to present increasing levels of hepatospecific ligands.

Hepatocytes cultured on Matrigel typically adopt a rounded morph-ology and undergo extensive aggregation, organizing themselves into 3-D spheroidal structures via active cell-mediated matrix reorganization [69, 76]. This morphogenesis may be due to the malleability of Matrigel in addition to the biochemical signaling of the matrix components it possesses. Previous studies have also demonstrated an ability to modulate hepatocyte morpho-genesis and aggregation by altering the mechanical compliance of Matrigel via chemical crosslinking [76]. Since Matrigel is a substrate with readily mod-ifiable mechanical properties as well as an intrinsically high level of bioadhe-sivity, it represents a useful model to investigate the sensitivity of hepatocyte morphogenesis and aggregation dynamics to a specific biophysical charac-teristic of the ECM. In addition to mechanical modification of the culture microenvironment, hepatocyte cultures were exposed to varying amounts of GF stimulation by the incorporation of the GFs, EGF and HGF at varying concentrations into the culture medium in order to dramatize the cellular responsiveness to differential degrees of substrate rigidity.

While the Matrigel culture system was developed to investigate the ef-fects of alterations in substrate mechanical compliance on ensuing hepa-tocellular behavior, the role of specific substrate chemistry in mediating this response cannot be isolated since the underlying Matrigel substrate is composed of a complex variety of bioactive molecules. As an alternative to the Matrigel system, the second hepatocellular culture system utilizes polyacrylamide-based hydrogels, which possess several significant advantages including a basal nonadhesivity to cells, precise, systematic control of substrate flexibility via the ratio of bisacrylamide crosslinker to acrylamide monomer present upon polymerization [123], and the ability to be derivatized with hetero-bifunctional crosslinking groups following gel polymerization that can subsequently be used to covalently couple bioactive ligands to the gel surface [124, 125]. Previously, polyacrylamide gels have been previously shown to support hepatocyte adhesion and viability when conjugated irreversibly to ligands such as hepatic lectins [126, 127], and a nearly 100-fold increase in elastic storage modulus has been previously reported in low wt% acrylamide gels with a 0.2% increase in the amount of crosslinking agent, a result that had considerable effects on the shape and elongation of cells grown on their surfaces [128, 129]. Thus, the second model system is designed to incorporate a well-defined substrate system whose hepatocellular adhesive properties can be precisely controlled by the substrate presentation of a specific bioactive ligand and whose mechanical properties can be altered over a similar range to the initial model system. Specifically, the second model system aimed to utilize varying degrees of polyacrylamide gel rigidity along with varying levels of a substrate-presented ECM ligand, fibronectin, in order to establish differential degrees of hepatocellular morphogenesis and thereby induce distinct levels of expression of liver-specific function [144]. These studies can highlight not only hepatocellular response to substrate rigidity but also mechanosensitive patterns of ligand-induced morphogenetic and functional signaling.

4.1.2
Preparation of Differentially Compliant Hydrogel Substrates

For the first culture system, freshly isolated rat hepatocytes were seeded into dishes coated with Matrigel, chemically crosslinked using glutaraldehyde to varying degrees of mechanical compliance. These gel substrates were prepared by evenly distributing $33.7\,\mu L/cm^2$ of BioMatrix (Biomedical Technologies, Stoughton, MA) into six-well plates and incubating them at $37\,^{\circ}C$ for approximately one hour. BioMatrix is a secretory extract from Engelbroth–Holm–Swarm sarcoma, whose composition (85% laminin; 5% heparan sulfate proteoglycan; 5% collagen IV; entactin; nidogen) and properties are similar to that of Matrigel (Collaborative Biomedical Products, Bedford, MA). This formulation of substrate is generically referred to henceforth as compliant Matrigel. In selected cases, rigid Matrigel was formulated via a crosslinking procedure involving the addition of glutaraldehyde (Fisher Scientific, PA) subsequent to gelation. After Matrigel substrates were allowed to gel for 1 hour at $37\,^{\circ}C$, 1 mL of 0.025% (v/v) glutaraldehyde in phosphate-buffered saline (PBS) was added to each well. After incubation for 2.5 minutes, the glutaraldehyde solution was removed, and each well was washed with 50 mM NH$_4$Cl (Fisher Scientific, Pittsburgh, PA) in order to quench free aldehyde groups then with unsupplemented Dulbecco's modified Eagle medium (DMEM) in order to ensure the complete removal of excess glutaraldehyde prior to cell seeding. After allowing the cells to attach for 30 minutes, medium was exchanged with supplemented DMEM containing varying amounts (0–20 ng/mL) of epidermal growth factor (EGF) and hepatocyte growth factor (HGF) (Sigma, St. Louis, MO). Medium was subsequently exchanged each day over the course of culture.

For the second differentially compliant culture system, polyacrylamide hydrogels were polymerized within glass molds separated by plastic spacers (Bio-Rad, Hercules, CA), subsequently cut and adhered to the bottoms of 96-well tissue-culture plates in an array-based format, and finally functionalized by covalent conjugation of fibronectin molecules. The degree of mechanical compliance of each set of hydrogels was controlled precisely by the ratio of bisacrylamide crosslinking monomers to acrylamide backbone monomers present during gel polymerization. More specifically, gels were prepared in 1-mm-thick molds from aqueous precursor solutions of 20% (w/v) of acrylamide monomer (Acros, Somerville, NJ) and 0.028–1.8% (w/v) of N',N'-methylene bisacrylamide monomer (Acros) using N,N,N',N'-tetramethylenediamine (TEMED) (Acros) as an accelerator and ammonium persulfate (APS) (Acros) as an initiator. Gel slabs, sufficiently washed to remove any unreacted monomers, were allowed to reach equilibrium swelling volumes and were subsequently cut into 0.25-inch discs using a hole punch, a diameter similar to that of a well inside a 96-well plate. These discs were then adhered to the bottoms of 96-well plates (Nunc Nalge International, Rochester, NY) using a drop of optically clear glue (Norland, Cranbury, NJ) and subsequent exposure to UV light.

Polyacrylamide gel discs affixed to 96-well plates were functionalized with ligands by adapting a protocol for preparing activated polyacrylamide sheets [124]. Here, ligand immobilization to polyacrylamide gel substrates was accomplished via the coupling of a hetero-bifunctional crosslinking group, Sulfo-SANPAH (Pierce Biochemicals, Rockford, IL), to the gel surface following gel polymerization that can subsequently be utilized to covalently bind the amino groups of proteins. To prepare gel surfaces for covalent ligand coupling, hydrogel discs were overlaid with 0.64 mM sulfo-SANPAH (Pierce Chemicals, Rockford, IL) and then irradiated with UV light. After this photoactivation procedure was repeated once to ensure sufficient derivatization of the gel surface, gels were washed and overlaid with solutions of rat fibronectin (Sigma, St. Louis, MO) over a wide range of concentrations (bulk concentrations of 0–20 pmol/cm^2). Following the conjugation of ligands to hydrogel surfaces, gels were neutralized via exposure to 1 M ethanolamine (Fisher), converting any remaining active esters within hydrogels to neutral amides. Subsequently, hydrogel discs were washed with a solution containing 1 M NaCl and 0.1 M KH_2PO_4 and with a 0.01 M KH_2PO_4 solution in order to remove any entrapped proteins from within the gel bulk. Prior to cell seeding, gels were blocked with 1% (w/v) bovine serum albumin (BSA, Sigma) and washed with sterile basal medium. In order to detect the presence of varying amounts of fibronectin immobilized on hydrogel surfaces, an ELISA was performed using a rabbit anti-rat antibody to fibronectin (Calbiochem, San Diego, CA) and an horseradish peroxidase (HRP)-conjugated goat anti-rabbit antibody to IgG (Sigma). Levels of fibronectin binding to activated hydrogels were shown to increase with bulk fibronectin concentration of a wide range of conditions and were relatively independent of the level of gel crosslinking.

4.1.3
Measurement of Substrate Stiffness via Rheometry

Rheological measurements of hydrogel substrates were performed in order to determine changes in mechanical properties following either chemical crosslinking after gelation (Matrigel) or variation in the concentration of crosslinking groups prior to polymerization (polyacrylamide gels). Here, the mechanical strength of gel substrates was probed using rheological measurements on a SR-2000 dynamic stress rheometer (Rheometrics, Piscataway, NJ). Both the elastic storage modulus (G') and the loss modulus (G'') were measured as a function of angular frequency during a dynamic frequency sweep. For small oscillatory shears, the shear modulus, G, also known as the modulus of rigidity is defined as:

$$G\left(t, \gamma_0\right) = \frac{\sigma(t)}{\gamma_0},$$

where $\sigma(t)$ is the shear stress at time t and γ_0 is the shear strain amplitude [130]. The shear stress function, $\sigma(t)$, is defined as:

$$\sigma(t) = \sigma_0 \sin(\omega t + \delta),$$

where σ_0 is the stress amplitude, ω is the frequency, and δ is the phase shift. The elastic storage, G', and loss, G'', moduli can be derived from the following equations:

$$\sigma(t) = \gamma_0 \left[G'(\omega) \sin \omega t + G''(\omega) \cos \omega t\right],$$
$$G_d = \frac{\sigma_0}{\gamma_0},$$
$$G' = G_d \cos(\delta),$$
$$G'' = G_d \sin(\delta).$$

For a purely elastic material (solid), there is nearly no viscous dissipation, which corresponds to a loss modulus of virtually zero. On the other hand, for a purely viscous liquid, there is no energy storage, corresponding to storage modulus values of zero and a loss angle of 90°. Thus, the magnitudes of the elastic moduli represent a measure of resistance to shear strain while the disparity between the elastic and storage moduli indicates by what means the shear strain is translated through the material.

Dynamic stress rheometry was performed to assess the mechanical strength of Matrigel substrates in a parallel-plate geometry using 40-mm plates. Compliant and rigid Matrigel samples were prepared as previously described on the lower plate of the rheometer. The lower plate was then heated to 37 °C (±1.0 °C), and a lid was placed over the sample to prevent drying. After allowing a period of 1 h for gelation, the lid was removed, and the upper plate was lowered to a gap height of 0.25 mm. The sample was then subjected to oscillatory shear at a constant stress of 0.2 Pa by rotation of the upper plate. Both the elastic storage modulus (G') and the loss modulus (G'') were measured as a function of angular frequency during a dynamic frequency sweep. For all Matrigel samples tested, the storage modulus (G') was invariably greater than the loss modulus (G'') (Fig. 12A). In addition, values of both moduli increased monotonically with increasing frequency. Values of storage and loss moduli for chemically crosslinked rigid Matrigel were found to be approximately ten times greater in magnitude than those for basal compliant Matrigel throughout the frequency range studied. These variations are indicative of significantly increased stiffness values for rigid Matrigel.

In a similar manner to that for Matrigel substrates, rheological measurements of polyacrylamide gels were performed to determine changes in mechanical properties following the variation in the amount of methylene

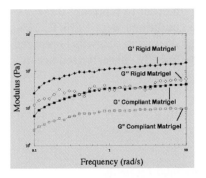
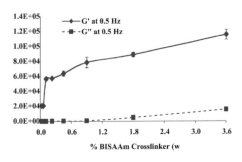

Fig. 12 A Mechanical properties of compliant and rigid Matrigel. Typical elastic response of each substrate to oscillatory shear at 0.2 Pa with a frequency sweep from 0.1 rad/s to 10 rad/s. **B** Mechanical properties of differentially crosslinked polyacrylamide gels containing 20% (w/v) AAm. Values of elastic moduli were recorded under oscillatory 2% strain at a frequency of 1 Hz. All measurements were averaged over at least three gel samples. (Figure 12A reproduced from Biotechnology and Bioengineering Vol. 69 (2000) with permission from Wiley Publishers)

Table 2 Characteristic properties of 0.25-inch-diameter gel discs prepared from 20% polyacrylamide gel slabs that had been polymerized with varying amounts of methylene-bisacrylamide crosslinker and were allowed to reach equilibrium swelling prior to gel sectioning. The original thickness of all gel slabs was set at 1 mm by mold-casting during polymerization. Water content was determined after two days of drying time under a laminar-flow hood

Amt. BISAAm crosslinker (%w/v)	Shear modulus G' (dynes/cm^2)	Equilibrium gel height (mm)	Water content (%)
1.80	90122 ± 2072	1.09 ± 0.01	74.60 ± 0.54
0.225	57080 ± 927	1.26 ± 0.01	89.79 ± 0.24
0.028	20310 ± 1087	1.48 ± 0.01	93.22 ± 0.24

bisacrylamide monomer in the polymer precursor solution (Fig. 12B). For these studies, hydrogel discs that were approximately 20 mm in diameter were cut using a cork borer, and rheometry was performed utilizing a 20-mm cone-and-plate geometry. Again, for all gels tested, the storage modulus (G') was invariably greater than the loss modulus (G''), and values of both moduli increased consistently with increasing crosslinking. Between the most loosely crosslinked condition [0.056% methylenebisacrylamide (BISAAm)] and the most heavily crosslinked condition (3.6% BISAAm) for 1-mm-thick 20% acrylamide (AAm) gels, over a fivefold difference in G' was detected using a dynamic strain-controlled frequency sweep at a constant 2% strain and a frequency of 1 Hz. It is also important to note here that, for ensuing experiments, three crosslinking conditions for 20% AAm gels were selected that possessed markedly different values of G' in order to represent three distinct levels of mechanical compliance. These conditions were 1.8%, 0.225%, and 0.028% BISAAm, and gel disc substrates (0.25 inch in diameter) prepared using these crosslinking conditions were subsequently characterized according to their equilibrium swelling properties (Table 2).

4.2
Results

4.2.1
Evaluation of Cellular Morphogenesis and Multicellular Assembly

In order to monitor cellular morphogenesis and aggregation, images were acquired using transmitted-light microscopy on either a Zeiss Axiovert microscope (Zeiss, Thornwood, NY) or an Olympus IMT-2 inverted microscope (Olympus, Lake Success, NY). For hepatocytes cultured on differentially compliant Matrigel substrates, markedly variant patterns of cellular morphogenesis that were dramatized by the presence of growth factors (GFs) evolved over the course of culture. In the absence of GFs, hepatocytes cultured on compliant Matrigel adopted rounded morphology, and, for the most part, exhibited minimal aggregation behavior (Fig. 13A). When cultures were incubated in the presence of either EGF or HGF, active aggregation was initiated by 24 h, which led to the formation of small spheroidal aggregates by 48 h (Fig. 13B;C). When cultures were co-stimulated with EGF and HGF, a marked increase in the number of active aggregates was evident by 24 h, resulting in the formation of large spheroidal aggregates by 48 h (Fig. 13D). Hepatocytes cultured on rigid Matrigel appeared more elongated in the absence of GFs, and aggregated significantly more than those on compliant Matrigel (Fig. 13E). Following EGF stimulation, hepatocytes on this substrate exhibited an elevated amount of spreading by 24 h, which led to a significant increase in multicellular aggregation by 48 h, characterized by large, semi-spheroidal aggregates (Fig. 13F). Stimulation with HGF further intensified the degree of hepatocyte 2-D spreading on rigid Matrigel compared to that observed for the EGF-stimulated condition. By 48 h, the degree of aggregation appeared similar to the EGF-stimulated condition, yet aggregate morphology was slightly more flattened and 2-D in nature (Fig. 13G). Simultaneous co-stimulation with EGF and HGF resulted in an even greater degree of spreading by 24 h than via HGF stimulation alone, and, by 48 h, this activation resulted in the formation of large, flattened, interconnected aggregate nodes bridged by cords of highly extended cellular processes (Fig. 13H). Following the first two days of culture, cell morphogenesis was observed to be nearly complete for all mechanochemical conditions.

In order to obtain quantitative measures of hepatocyte morphology on Matrigel substrates, single cell area (Fig. 15A) and cell aspect ratio (ratio of length of major axis to the length of the minor axis; data not shown) were computed after 48 h of culture. Together, these measurements of cell shape provide a characterization of the degree of cell spreading and elongation, respectively. Both single cell area and aspect ratio followed similar trends indicating that cell spreading was nonuniform and dominated by unidirectional elongation. On compliant Matrigel, there were insignificant changes in both cell area and aspect ratio between GF-deficient and GF-stimulated cultures, as cells remained approximately spheroidal in shape. However, on rigid Matrigel, GF stimulation resulted in increased levels of these parameters relative to those computed for GF-deficient cultures on both compliant and rigid Matrigel as well as GF-containing conditions on compliant Matrigel. These results supported the observed increases in 2-D cell spreading and elongation that were recorded for GF-stimulated cells on rigid Matrigel.

Hepatocellular morphogenesis on differentially compliant polyacrylamide gel substrates was determined to be critically governed by both gel mechanical compliance and fibronectin (FN) concentration. Notably, by 16 h of culture, cellular spreading was observed to be generally enhanced with increasing sub-

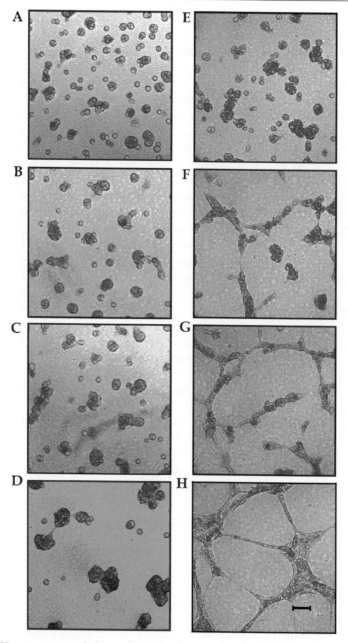

Fig. 13 Hepatocyte morphology after 48 h of culture on compliant Matrigel (**A–D**) and rigid Matrigel (**E–H**) under various activation conditions: **A,E** absence of GF stimulation; **B,F** 20 ng/mL EGF; **C,G** 20 ng/mL HGF; and **D,H** 20 ng/mL EGF and 20 ng/mL HGF. The *scale bar* represents 100 µm

strate rigidity as well as increasing FN concentration, although the effects of substrate mechanical compliance appeared to be more dominant (images not shown). In particular, in a manner independent of FN concentration, the most compliant gels did not engender any significant cellular spreading. Hepatocytes cultured on gels of high and intermediate rigidity exhibited high degrees of cell spreading for FN concentrations of 2 pmol/cm^2 or above, while the degree of cellular spreading at 0.2 pmol/cm^2 was less pronounced on both substrates, especially on gels of intermediate levels of compliance. It is also

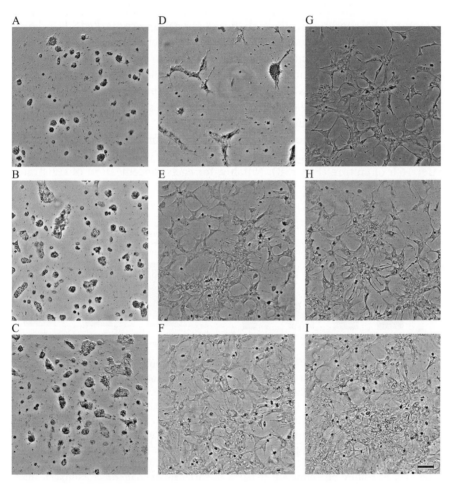

Fig. 14 Transmitted-light images depicting hepatocyte morphology after three days of culture on 20% AAm polyacrylamide hydrogels of varying compliance and ECM ligand presentation. The amount of BISAAm crosslinker (% w/v) ranged between 0.028% (**A–C**) to 0.225% (**D–F**) to 1.8% (**G–I**) and the concentration of substrate-immobilized fibronectin (FN) was varied from 0.2 (**A,D,G**) to 2 (**B,E,H**) to 20 pmol/cm^2 (**C,F,I**). Images were captured at a magnification of 10×. *Scale bar* represents 100 μm

important to note that, at this concentration for all gel types, there was a noticeable reduction in the attachment efficiency of hepatocytes to the gel, perhaps
due to the nonadhesive nature of polyacrylamide gels in their nonfunctional-
ized form to cells. By the third day of culture, initial cellular morphogenesis
evolved into either 2-D or 3-D aggregation in a manner dependent on the
mechanochemical characteristics of the substrate. On the most compliant substrates (Fig. 14A–C), cells and cell aggregates remained fairly spheroidal in
shape in all culture conditions, although increasing substrate-bound FN concentration did enhance the average size of spheroidal aggregates. On the most
rigid gels (Fig. 14G-I), hepatocytes formed 2-D highly interconnected aggregates for all FN conditions, although these aggregates were markedly less dense
for low-FN conditions. On gels of intermediate mechanical compliance, cells
also formed 2-D highly interconnected multicellular structures for high- and
intermediate-FN conditions (Fig. 14E,F); however, hepatocytes at the lowest-
FN condition exhibited a markedly different morphology (Fig. 14D). Here,
although fewer cells were clearly present, aggregates were dramatically more
three-dimensional in nature.

Hepatocellular morphogenesis was quantified for cultures on functionalized differentially compliant polyacrylamide gels by evaluating single cell area
after 16 h of culture (Fig. 15B). Notably, for each concentration of substrate-

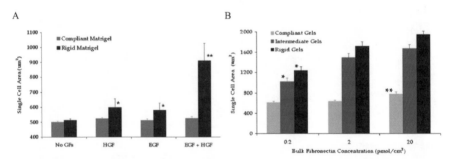

Fig. 15 A Substrate-specific effects of time-invariant GF stimulation on hepatocyte morphology (single-cell area) after 48 h of culture on compliant and rigid Matrigel. *Single
asterisks* represent statistically significant differences ($p < 0.05$) between morphology indices computed from single GF-stimulated cultured and GF-deficient cultures on rigid
Matrigel. The *double asterisk* indicates statistical significance between hepatocyte cultures
under GF co-stimulatory conditions on rigid Matrigel and all other conditions. **B** Effects
of both polyacrylamide gel rigidity and concentration of the substrate-immobilized adhesive ECM ligand, fibronectin, on hepatocyte morphology. Single cell area was computed
after 16 h of culture on compliant (0.028% methylenebisacrylamide crosslinker), intermediate (0.225% crosslinker), and rigid (1.8% crosslinker) polyacrylamide gels. *Single
asterisks* indicate statistical significance of morphology indices between measurements
from the lowest FN cultures and all higher FN cultures at a given substrate rigidity.
The *double asterisk* represents statistically significant differences between the highest FN
cultures and all lower FN cultures at a given substrate rigidity. All *error bars* indicate
standard errors around the mean

immobilized fibronectin, cellular spreading increased with decreasing substrate compliance. These differences were measured to be the most dramatic between the most compliant gels and gels of intermediate compliance, where, in certain cases, a nearly three-fold increase was detected. In addition, increasing amounts of substrate-presented fibronectin also caused considerable intensification of cellular spreading; a result that supported the premise that fibronectin can be a strong hepatocellular morphogen. This phenomenon was most prominent for cultures of intermediate through low fibronectin exposure ($2-0.2$ pmol/cm^2) for cultures on gels of intermediate or low compliance engendering as much as a 40–47% increase in cell spreading. For hepatocyte cultures on compliant gels, this effect was much less pronounced supporting the observations that cells were incapable of significant spreading on these surfaces.

4.2.2
Expression of Liver-Specific Function

For hepatocytes cultured on compliant and rigid Matrigel under various growth factor conditions, values of albumin secretion rates were computed for days 3–4 of culture (data not shown) and days 6–7 of culture (Fig. 16A). Here, albumin data was normalized to the numbers of cells seeded since cellular attachment to Matrigel, whether in compliant or rigid form, was nearly 100%, and no significant cell detachment was detected over the culture period (data not shown). On compliant Matrigel by day 4, significant differences were detectable in the levels of albumin secretion on co-stimulated, single GF-stimulated, and GF-deficient conditions. In particular, the presence of either GF enhanced albumin secretion rate, while GF co-stimulation produced the most pronounced increase in albumin secretion rate, nearly three-fold versus GF-deficient cultures. By day 7, functional levels of HGF-stimulated hepatocytes were somewhat lower than that of EGF-stimulated hepatocytes, although functions were consistently higher in both cultures than those measured in GF-deficient cultures and lower than those measured in GF co-stimulated cultures. On rigid Matrigel, GF co-stimulation resulted in a sharp decline in albumin secretion rates relative to levels measured on compliant Matrigel at both days 3–4 and days 6–7. In addition, these levels were lower than those recorded for stimulation with EGF, a condition that resulted in the greatest amount of albumin secretory activity. GF-deficient cultures on rigid Matrigel still produced lower albumin expression than stimulation by either GF, yet they elicited equivalent rates of albumin secretion as GF co-stimulation.

For differentially compliant polyacrylamide gel cultures, albumin secretion was measured following the third and fifth day of culture in order to evaluate the expression of hepatocellular differentiated function. All data was normalized on a per-cell basis to the numbers of adherent cells as determined by

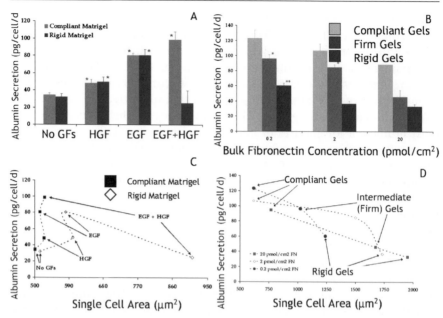

Fig. 16 A Comparative effects of GF stimulation on long-term (day 6 + 7) albumin secretion rates between hepatocytes on compliant and rigid Matrigel. All concentrations of GFs were 20 ng/mL. *Single asterisks* represent statistical significance ($p < 0.05$) for data relative to that for GF-deficient conditions. **B** Effects of substrate adhesivity, in terms of substrate-presented fibronectin (FN), on hepatocellular albumin secretion rates (day 3 + 5) on compliant (0.028% methylenebisacrylamide crosslinker), intermediate (0.225% crosslinker), and rigid (1.8% crosslinker) polyacrylamide gels. Single asterisks represent statistical significance ($p < 0.05$) for data relative to that for the most highly adhesive conditions (20 pmol/cm^2 FN) at a given substrate rigidity. *Double asterisks* represent statistical significance for data relative both moderately (2 pmol/cm^2 FN) and highly adhesive (20 pmol/cm^2 FN) conditions at a given substrate rigidity. **C,D** Relationship between single-cell area and albumin secretion rates for hepatocytes cultured on differentially compliant Matrigel substrates (**C**) and polyacrylamide gels (**D**)

Calcein AM staining (Molecular Probes, Eugene, OR) using a Cytofluor plate reader (Applied Biosystems, Foster City, CA) and comparison to collagen gel controls, where cell attachment remained at nearly 100% levels throughout the culture period. For both day 3 and day 5 cultures, similar trends in albumin secretion rates were detected. The averages of these results are plotted in Fig. 16B. In particular, increasing substrate rigidity led to a reduction in albumin secretion rates independent of the amount of substrate-presented fibronectin (FN). In this manner, the cells cultured on the most compliant gels consistently expressed the highest levels of albumin secretory activity. Additionally, increasing levels of substrate-immobilized FN also led to a marked reduction in hepatocellular function, consistent with previous reports that FN is a growth-inductive ligand (thereby leading to reduced function by a reciprocal balance) [80, 131]. These effects were the most dramatic for cultures on rigid gels between low FN exposure ($0.2\,pmol/cm^2$) and intermediate FN exposure ($2\,pmol/cm^2$) as well as on intermediate gels between high FN exposure ($20\,pmol/cm^2$) and intermediate FN exposure ($2\,pmol/cm^2$). Interestingly, on compliant gel substrates, a high level of albumin secretion was maintained even at the highest degree of FN presentation suggesting that the differentiation-reducing capacity of FN may be suppressed at this malleable level of substrate mechanics.

4.3
Conclusions

4.3.1
Substrate Stiffness Determines the Outcome of Active Cellular Morphogenesis

Cellular morphogenesis is thought to be the result of both cytoskeleton-generated tractile forces and the resistance offered at the specific attachment sites by the surrounding matrix, which is governed by its stiffness [132–134]. Here, in these studies, it is hypothesized that, for cultures on compliant substrates, while active hepatocyte spreading is inhibited by the malleability of the matrix, the exertion of cellular tractile forces is significant enough to re-organize the gel, thus leading to 3-D aggregation of relatively compacted cells. Previous studies have similarly reported that matrix reorganization correlates strongly with the directionality and extent of hepatocyte aggregation on basal Matrigel [76]. On the other hand, on more rigid substrates, the stiff substrate may provide sufficient resistance to cell tractile forces, resulting in cell spreading, elongation, and increased 2-D interconnectivity between aggregates. The studies highlighted in this review support these hypotheses by demonstrating not only how substrate rigidity is remarkably critical to this process, but also how cellular activation (via GF stimulation) and the number of morphogenetically active adhesive contacts (via the presence of substrate-immobilized FN) can affect the outcome of cytoskeletal rearrangement.

For both the Matrigel and polyacrylamide gel systems, changes in substrate mechanical compliance strongly limited resultant cellular geometry as well as the type of cell–cell interactions that could evolve during cellular aggregation. In particular, for studies using Matrigel, cellular spreading and elongation were significantly elevated with increasing substrate stiffness. In the presence of growth factors, changes in morphogenesis were especially dramatic between compliant and rigid Matrigel as the stiffer, crosslinked Matrigel supported high levels of 2-D aggregation relative to the 3-D spheroids present on highly malleable, basal Matrigel. Even in the absence of GFs, the rigid substrate induced subtle increases in both cell area and elongation compared to compliant Matrigel. While the rigid substrate may be more suited to support aggregation behavior independent of GF stimulation, GFs may be responsible for dramatizing patterns of cellular morphogenesis by inducing increased generation of cellular tractile force, via the upregulation of molecular motor-encoding genes. In previous reports, HGF, along with EGF, was shown to potent in cytoskeletal gene upregulation [135] as well as eliciting cell elongation and cell "scattering" on deformable collagen gels [136] and Matrigel-coated surfaces [137]. As GFs, EGF and HGF serve as model factors for cooperative effects, given that these GFs bind to distinct receptors (EGF and c-met, respectively) [138–140], and activate cells by eliciting distinct signal transduction pathways [137, 141]. Supporting this theory, GF co-stimulation was responsible for the largest differences in cell morphogenesis observed between compliant and rigid Matrigel substrates leading to a greater degree of cell aggregation than that elicited via stimulation by either EGF or HGF presented alone, or by either GF at twice the dosage concentration (data not shown).

For polyacrylamide gel cultures, the varying concentrations of substrate-immobilized FN were critical for understanding how substrate mechanical compliance and the extent of morphogenetically active cell–substrate contact may both play roles in determining hepatocellular morphogenesis [144]. This relationship was particularly apparent for hepatocytes cultured on gels of intermediate compliance, as cell morphogenesis was the most sensitive to FN concentration on these substrates. Here, while the substrate was sufficiently rigid to enable cell spreading and 2-D aggregation under conditions of high levels of FN presentation ($\geq 2 \, pmol/cm^2$), these cellular processes were much less pronounced at low degrees of substrate-bound FN, and some 3-D semi-spheroidal aggregates were able to form. On the contrary, for highly rigid substrates, the presence of FN only affected the efficiency of cellular attachment and the kinetics of cell morphogenesis as cells spread markedly faster in the presence of increasing levels of FN. However, morphogenesis eventually resulted in similarly organized 2-D interconnected aggregates independent of FN concentration. Likewise, for highly compliant substrates, the presence of FN influenced the efficiency of cellular attachment as well as the kinetics of cellular morphogenesis, and, although the result of cell morphogenesis was quite different than that for highly rigid substrates (as hepatocytes on compliant gels formed multicellular spheroidal aggregates), cells and cell aggregates had a similar final morphology that was independent of FN concentration.

These observations suggest that, at certain levels of substrate rigidity (intermediate compliance), the number of morphogenetically active cell–substrate attachment sites may be more important to the outcome of cellular morphogenesis that at other levels (low and high compliance), where the mechanical resistivity of the substrate alone can dominant its progression. For cultures on gels of intermediate compliance, where the substrate is neither completely rigid (providing complete resistance to tractile forces and forcing spreading onto cells that wish to maximize cell-substrate adhesion) nor completely compliant (providing little resistance to tractile forces and causing cells to remain spheroidal in shape independent of the number of cell–matrix contacts), a combination of FN-mediated cell–substrate adhesions and underlying gel rigidity may underlie the observed spreading behavior. At high FN concentrations on this substrate, FN, a noted hepatocyte morphogen, is able to induce spreading and 2-D reorganization by not only providing anchors for which cells can exert tractile forces but also by stimulating cells for morphogenesis. However, at low FN concentrations on this substrate, the lack of a sufficient morphogenetic signal along with marginal substrate resistivity causes cells to exhibit far less spreading behavior and forces aggregates to be more 3-D in nature. Here, hepatocytes, under these conditions, may have a greater affinity for each other than for the weakly functionalized substrate after intercellular contacts are formed and, unlike cultures on rigid gels, the intermediate compliance of the substrate allows for this 3-D rearrangement to occur.

4.3.2
Mechanochemically Induced Morphologic State Regulates Functional Fate

In these studies, combinations of mechanical and chemical variations in the culture environment resulted in not only distinct patterns of cellular morphogenesis but also distinct cell growth and liver-specific functional responses of hepatocytes. These patterns of hepatocellular activity were highly correlated to cell morphologic state for both cultures on Matrigel (Fig. 16C) and FN-functionalized polyacrylamide gel substrates (Fig. 16D), as increased cellular spreading was typically associated with decreased functional expression. For Matrigel cultures, this relationship was most apparent under conditions of GF co-stimulation for rigid substrates, where the only dramatic increases in cellular spreading and elongation were detected. Here, function was drastically reduced in comparison to all other conditions, where the levels of GF stimulation appear to have a much larger role than small variations in cell morphology on maintaining hepatospecific function. Using a strategy that incorporated dynamic GF stimulation on differentially compliant Matrigel substrates, the relative roles of cell morphologic state and GF activation were decoupled in a subsequent study, which concluded that cell morphology had the ability to amplify or restrict the functional responsiveness of hepatocytes to GFs [142].

For polyacrylamide gel cultures, a much wider range of cellular morphogenesis was obtained via simultaneous manipulation of substrate mechanics and adhesivity. Thus, the relationship between cellular spreading and function could be examined for a number of morphologic states and not only the extreme ones observed in the Matrigel system. This data provided strong evidence that function was highly correlated to cellular shape; as for the conditions examined in the polyacrylamide gel system, cell function was nearly always reduced with increasing cell area, independent of FN exposure. The exceptions to this trend did establish a secondary function-reducing effect of FN on cellular functional fate. Interestingly, the correlation between cell shape and function was approximately linear over the range of compliance conditions studied for the highest level of FN presentation, but became markedly less linear and shifted to the left for lower degrees of FN presentation. This result may be due to an incomplete level of adhesion by cells to the substrate under these conditions of limited ligand availability, which, in turn, limits the degree to which cells can spread.

The role of cell–cell interactions appeared to be secondary to cell morphology in determining the level of functional activity in both culture systems. It has been documented that cell–cell contact may also not be essential for maintenance of hepatocyte functions, particularly on soft, compliant substrates such as basal Matrigel [69]. Similarly, from the present study on substrates with varying stiffness, no universal correlation between the extent of hepatocyte aggregation and resultant liver-specific function was observed. In fact, the highest levels of aggregation (indicative of high cell–cell contact) which were achieved following either GF co-stimulation on rigid Matrigel or high levels of FN presentation on rigid polyacrylamide gels, were also accompanied by the lowest levels of function, suggesting that the extent of cell–cell contact is likely not to be a primary determinant of function under these conditions. It is not clear from the results of these studies whether, alternatively, the geometric nature of cell–cell contact (2-D or 3-D) may also play a critical role in modulating functional expression. As an alternative to the possible role of cell–cell contact, it is believed that changes in the cellular phenotype may critically govern hepatocellular growth and function on differentially compliant substrate systems. In particular, these phenomena may partly explain the inverse aggregation-function relation elicited in hepatocyte cultures exhibiting high degrees of 2-D aggregation.

5
Summary

Overall, the studies highlighted in this review demonstrate the potent ability of substrate biophysics to regulate hepatocellular morphogenesis and reorganization and, subsequently, modulate cell functional fate via morphologic priming. Substrate microtopography and substrate mechanical compliance thus represent distinct tools that can be utilized as design parameters in the development of hepatospecific constructs, as each of these two biophysical properties was shown to affect hepatocellular phenotype significantly. Cell morphogenesis was also demonstrated to be a coordinated response to both substrate biophysics and various forms of biochemical activation, either adhesive or soluble. In this manner, similar states of substrate biophysics could engender diverse morphogenetic responses by either providing additional traction (increasing adhesivity) or additional tractile force (increasing GF activation). The expression of liver-specific function was also determined to be highly correlated to the degree of biophysically regulated cell morphogenesis. For all of the cases examined in this review, increased cellular spreading was strongly linked to a reduction in hepatocellular functional activity, yet the nature of this relationship was also highly sensitive to the levels of substrate bioactivity and soluble stimulation. A cartoon schematic summarizing the nature of this correlation is presented in Fig. 17. Notably, in this diagram, substrate bioactivity and soluble stimulation can shift the curve relating cell function to cell shape. The shift intensity may also be dependent on the morphologic state of the cells, thereby amplifying or restricting the biochemical effects, particularly at levels of intermediate cell spreading. These conclusions are supported by studies by Moghe and coworkers, which attempt to decouple the role of cell morphology from that of GF activation in governing the expression of hepatocellular function [142].

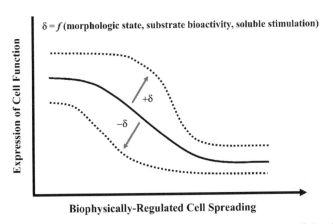

Fig. 17 Cartoon schematic depicting the correlation between hepatocellular functional output and cell phenotype. Cell morphogenesis is regulated by substrate biophysical properties as well as being driven by cellular adhesivity and the exertion of cell forces. In addition to the role of cellular morphologic state, the degree of cell activation due to either substrate-immobilized or soluble ligand-specific interactions may also play a key role in determining cell functional fate, possessing the ability to further shift the curve relating cell function to cell shape (*from the solid line to the dotted lines*). The amount of this shift, δ, is thus believed to be a function of both substrate bioactivity and soluble stimulation and may also be dependent on the morphologic state. Here, certain biophysically induced cell morphologies may be more elastic than others that may be more restrictive to functional enhancement or repression, particularly at the extremes of cell morphology

By understanding the nature through which the biophysical and biochemical attributes of the culture microenvironment act in combination towards determining hepatocellular response, researchers can not only manipulate cell activity in vitro, but also provide a basis for the strategic design of implantable cell scaffolds, which need to be capable of inducing, regulating, and sustaining specific levels of hepatocellular function. In this manner, the ultimate ability to harness the dramatic sensitivity of hepatocyte activity to the mechanochemical properties of the culture environment may engender the successful development of cell-seeded biomaterial constructs as tissue-substitute devices, which can be precisely attuned to meet particular clinical requirements during potential therapies for liver failure.

Acknowledgements This work was supported in part by: NIH NIBIB P41 grant (EB000922-01), NSF CAREER grant (BES9733007); Merck Exploratory Research Award; Johnson & Johnson Discovery Award; the Rutgers SROA Award; and the NIH-sponsored Rutgers UMDNJ Biotechnology Training Program. Technical help from Perry Lancin, Thomas Brieva, and Anouska Dasgupta, and laboratory access by Professor Frederick Kauffman are gratefully acknowledged.

References

1. Alenghat FJ, Ingber DE (2002) Sci STKE 2002:E6
2. Curtis A, Riehle M (2001) Phys Med Biol 46:R47
3. Katz BZ, Zamir E, Bershadsky A, Kam Z, Yamada KM, Geiger B (2000) Mol Biol Cell 11:1047
4. Discher DE, Janmey PA, Wang YL (2005) Science 310:1139
5. Davis GE, Bayless KJ, Mavila A (2002) Anat Rec 268:252
6. Geiger B, Bershadsky A, Pankov R, Yamada KM (2001) Nat Rev Mol Cell Biol 2:793
7. Huang S, Ingber DE (2000) Exp Cell Res 261:91
8. Hodgkinson CP, Wright MC, Paine AJ (2000) Mol Pharmacol 58:976
9. Sawamoto K, Takahashi N (1997) In Vitro Cell Dev Biol Anim 33:569
10. Hamamoto R, Yamada K, Kamihira M, Iijima S (1998) J Biochem (Tokyo) 124:972
11. Hansen LK, Hsiao C, Friend JR, Wu FJ, Bridge GA, Remmel RP, Cerra FB, Hu W (1998) Tissue Eng 4:65
12. Moghe PV, Berthiaume F, Ezzell RM, Toner M, Tompkins RG, Yarmush ML (1996) Biomaterials 17:373
13. Singhvi R, Kumar A, Lopez GP, Stephanopoulos GN, Wang DIC, Whitesides GM, Ingber DE (1994) Science 264:696
14. Dispersio CM, Jackson DA, Zaret KS (1991) Mol Cell Biol 11:4405
15. Tilles AW, Berthiaume F, Yarmush ML, Tompkins RG, Toner M (2002) J Hepatobiliary Pancreat Surg 9:686
16. Hasirci V, Berthiaume F, Bondre SP, Gresser JD, Trantolo DJ, Toner M, Wise DL (2001) Tissue Eng 7:385
17. Kneser U, Kaufmann PM, Fiegel HC, Pollok JM, Kluth D, Herbst H, Rogiers X (1999) J Biomed Mater Res 47:494
18. Sugimoto S, Mitaka T, Ikeda S, Harada K, Ikai I, Yamaoka Y, Mochizuki Y (2002) J Cell Biochem 87:16

19. Powers MJ, Domansky K, Kaazempur-Mofrad MR, Kalezi A, Capitano A, Upadhyaya A, Kurzawski P, Wack KE, Stolz DB, Kamm R, Griffith LG (2002) Biotechnol Bioeng 78:257
20. Glicklis R, Shapiro L, Agbaria R, Merchuk JC, Cohen S (2000) Biotechnol Bioeng 67:344
21. Karamuk E, Mayer J, Wintermantel E, Akaike T (1999) Artif Organs 23:881
22. Puviani AC, Lodi A, Tassinari B, Ottolenghi C, Ganzerli S, Ricci D, Pazzi P, Morsiani E (1999) Int J Artif Organs 22:778
23. Ranucci CS, Moghe PV (1999) Tissue Eng 5:407
24. Ranucci CS, Kumar A, Batra SP, Moghe PV (2000) Biomaterials 21:783
25. Semler EJ, Lancin P, Moghe PV (2003) in preparation
26. Semler EJ, Ranucci CS, Moghe PV (2000) Biotechnol Bioeng 69:359
27. American Liver Foundation (2001) Your Liver, A Vital Organ Retrieved Dec. 7 2001 http://www.liverfoundation.org/html/livheal_nu.htm
28. American Liver Foundation (2001) Fact Sheet: Hepatitis, Liver, And Gallbladder Diseases in the United States Retrieved Dec. 7 2001 http://www.liverfoundation.org/html/livheal_nu.htm
29. Diehl AM (2002) Front Biosci 7:e301
30. Court FG, Wemyss-Holden SA, Dennison AR, Maddern GJ (2002) Br J Surg 89:1089
31. Krasko A, Deshpande K, Bonvino S (2003) Crit Care Clin 19:155
32. Wiesner RH, Rakela J, Ishitani MB, Mulligan DC, Spivey JR, Steers JL, Krom RA (2003) Mayo Clin Proc 78:197
33. Thalheimer U, Capra F (2002) Dig Dis Sci 47:945
34. Davis MW, Vacanti JP (1996) Biomaterials 17:365
35. Jasmund I, Bader A (2002) Adv Biochem Eng Biotechnol 74:99
36. Strom S, Fisher R (2003) Gastroenterology 124:568
37. Nagamori S, Hasumura S, Matsuura T, Aizaki H, Kawada M (2000) J Gastroenterol 35:493
38. Tilles AW, Berthiaume F, Yarmush ML, Toner M (2002) Technol Health Care 10:177
39. Ohashi K, Park F, Kay MA (2001) J Mol Med 79:617
40. Miyashita H, Suzuki A, Fukao K, Nakauchi H, Taniguchi H (2002) Cell Transplant 11:429
41. Austin TW, Lagasse E (2003) Mech Dev 120:131
42. Rambhatla L, Chiu CP, Kundu P, Peng Y, Carpenter MK (2003) Cell Transplant 12:1
43. Tzanakakis ES, Hess DJ, Sielaff TD, Hu WS (2000) Annu Rev Biomed Eng 2:607
44. Allen JW, Hassanein T, Bhatia SN (2001) Hepatology 34:447
45. Allen JW, Bhatia SN (2002) Tissue Eng 8:725
46. Strain AJ, Neuberger JM (2002) Science 295:1005
47. Ochoa ER, Vacanti JP (2002) Ann NY Acad Sci 979:10
48. Babensee JE, McIntire LV, Mikos AG (2000) Pharm Res 17:497
49. Bhatia SN, Balis UJ, Yarmush ML, Toner M (1998) Biotechnol Prog 14:378
50. Bhatia SN, Balis UJ, Yarmush ML, Toner M (1998) J Biomater Sci Polym Ed 9:1137
51. Rana B, Mischoulon D, Xie Y, Bucher NL, Farmer SR (1994) Mol Cell Biol 14:5858
52. Ingber DE (2003) J Cell Sci 116:1157
53. Ingber DE (2002) Circ Res 91:877
54. Ryan PL, Foty RA, Kohn J, Steinberg MS (2001) Proc Natl Acad Sci USA 98:4323
55. Jalali S, del Pozo MA, Chen K, Miao H, Li Y, Schwartz MA, Shyy JY, Chien S (2001) Proc Natl Acad Sci USA 98:1042
56. Ingber DE (1997) Annu Rev Physiol 59:575

57. Wang N, Naruse K, Stamenovic D, Fredberg JJ, Mijailovich SM, Tolic-Norrelykke IM, Polte T, Mannix R, Ingber DE (2001) Proc Natl Acad Sci USA 98:7765
58. Ingber DE (2003) J Cell Sci 116:1397
59. Riveline D, Zamir E, Balaban NQ, Schwarz US, Ishizaki T, Narumiya S, Kam Z, Geiger B, Bershadsky AD (2001) J Cell Biol 153:1175
60. Balaban NQ, Schwarz US, Riveline D, Goichberg P, Tzur G, Sabanay I, Mahalu D, Safran S, Bershadsky A, Addadi L, Geiger B (2001) Nat Cell Biol 3:466
61. Hamill OP, Martinac B (2001) Physiol Rev 81:685
62. Plopper GE, McNamee HP, Dike LE, Bojanowski K, Ingber DE (1995) Mol Biol Cell 6:1349
63. Ingber DE (1997) Gravit Space Biol Bull 10:49
64. Popper HP (1957) Liver: structure and function. Plenum, New York
65. Michalopoulos GK, DeFrances MC (1997) Science 276:60
66. Zimmermann A (2002) Med Sci Monit 8:RA53
67. Nishio T, Iimuro Y, Nitta T, Harada N, Yoshida M, Hirose T, Yamamoto N, Morimoto T, Brenner DA, Yamaoka Y (2003) J Hepatol 38:468
68. Bucher NLR, Robinson GS, Farmer SR (1990) Seminars in Liver Disease 10:11
69. Moghe PV, Coger RN, Toner M, Yarmush ML (1997) Biotech Bioeng 56:706
70. Nagaki M, Shidoji Y, Yamada Y, Sugiyama A, Tanaka M, Akaike T, Ohnishi H, Moriwaki H, Muto Y (1995) Biochem Biophys Res Commun 210:38
71. Dunn JC, Tompkins RG, Yarmush ML (1991) Biotechnol Prog 7:237
72. Berthiaume F, Moghe PV, Toner M, Yarmush ML (1996) Faseb J 10:1471
73. Lindblad WJ, Schuetz EG, Redford KS, Guzelian PS (1991) Hepatology 13:282
74. Gardner MJ, Fletcher K, Pogson CI, Strain AJ (1996) Biochem Biophys Res Commun 228:238
75. Nagaki M, Sugiyama A, Naiki T, Ohsawa Y, Moriwaki H (2000) J Hepatol 32:488
76. Coger R, Toner M, Moghe P, Ezzell RM, Yarmush ML (1997) Tissue Eng 3:375
77. Singhvi R, Stephanopoulos G, Wang DIC (1994) Biotechnol Bioeng 43:764
78. Huang S, Chen CS, Ingber DE (1998) Mol Biol Cell 9:3179
79. Hamilton GA, Jolley SL, Gilbert D, Coon DJ, Barros S, LeCluyse EL (2001) Cell Tissue Res 306:85
80. Hansen LK, Mooney DJ, Vacanti JP, Ingber DE (1994) Mol Biol Cell 5:967
81. Hamada T, Kato Y, Terasaki T, Sugiyama Y (1997) J Hepatol 26:353
82. Takehara T, Matsumoto K, Nakamura T (1992) J Biochem 112:330
83. Parsons-Wingerter PA, Saltzman WM (1993) Biotechnol Prog 9:600
84. Powers MJ, Rodriguez RE, Griffith LG (1997) Biotechnol Bioeng 53:415
85. Krasteva N, Groth TH, Fey-Lamprecht F, Altankov G (2001) J Biomater Sci Polym Ed 12:613
86. Griffiths BJ, Evans PJ (2001) J Struct Biol 134:67
87. Bhadriraju K, Hansen LK (2002) Exp Cell Res 278:92
88. Ijima H, Nakazawa K, Mizumoto H, Matsushita T, Funatsu K (1998) J Biomater Sci Polym Ed 9:765
89. Hsiao CC, Wu JR, Wu FJ, Ko WJ, Remmel RP, Hu WS (1999) Tissue Eng 5:207
90. Weiss TS, Jahn B, Cetto M, Jauch KW, Thasler WE (2002) Cell Prolif 35:257
91. Ezzell RM, Toner M, Hendricks K, Dunn JC, Tompkins RG, Yarmush ML (1993) Exp Cell Res 208:442
92. Yamashita Y, Shimada M, Tsujita E, Shirabe K, Ijima H, Nakazawa K, Sakiyama R, Fukuda J, Funatsu K, Sugimachi K (2002) Cell Transplant 11:379
93. Mayer J, Karamuk E, Akaike T, Wintermantel E (2000) J Control Release 64:81
94. Catapano G, De Bartolo L, Vico V, Ambrosio L (2001) Biomaterials 22:659

95. Gomez-Lechon MJ, Jover R, Donato T, Ponsoda X, Rodriguez C, Stenzel KG, Klocke R, Paul D, Guillen I, Bort R, Castell JV (1998) J Cell Physiol 177:553
96. Hamazaki K, Doi Y, Koide N (2002) Hepatogastroenterology 49:1514
97. Chia SM, Wan AC, Quek CH, Mao HQ, Xu X, Shen L, Ng ML, Leong KW, Yu H (2002) Biomaterials 23:849
98. Hong JT, Glauert HP (2000) Toxicol In Vitro 14:177
99. Griffith LG, Lopina S (1998) Biomaterials 19:979
100. Yannas IV, Burke JF (1980) J Biomed Mater Res 14:65
101. Yannas IV, Burke JF, Gordon PL, Huang C, Rubenstein RH (1980) J Biomed Mater Res 14:107
102. Dagalakis N, Flink J, Stasikelis P, Burke JF, Yannas IV (1980) J Biomed Mater Res 14:511
103. Schoeters G, Leppens H, Van Gorp U, Van Den Heuvel R (1992) Cell Prolif 25:587
104. Horiguchi Y, Maruguchi T, Maruguchi Y, Suzuki S, Fine JD, Leigh IM, Yoshiki T, Ueda M, Toda KI, Isshiki N, et al. (1994) Arch Dermatol Res 286:53
105. Ohbayashi K, Inoue HK, Awaya A, Kobayashi S, Kohga H, Nakamura M, Ohye C (1996) Neurol Med Chir (Tokyo) 36:428
106. Nehrer S, Breinan HA, Ramappa A, Young G, Shortkroff S, Louie LK, Sledge CB, Yannas IV, Spector M (1997) Biomaterials 18:769
107. Saito M, Sakamoto T, Fujimaki M, Tsukada K, Honda T, Nozaki M (2000) Surg Today 30:606
108. Wintermantel E, Cima L, Schloo B, Langer R (1992) Exs 61:331
109. Mooney DJ, Park S, Kaufmann PM, Sano K, McNamara K, Vacanti JP, Langer R (1995) J Biomed Mater Res 29:959
110. Mooney DJ, Sano K, Kaufmann PM, Majahod K, Schloo B, Vacanti JP, Langer R (1997) J Biomed Mater Res 37:413
111. Nam YS, Park TG (1999) Biomaterials 20:1783
112. Semler EJ, Tjia JS, Moghe PV (1997) Biotechnol Prog 13:630
113. Sterzel W, Bedford P, Eisenbrand G (1985) Anal Biochem 147:462
114. Ben-Ze'ev A, Robinson GS, Bucher NLR, Farmer S (1988) Proc Natl Acad Sci USA 85:2161
115. von Recum AF, Shannon CE, Cannon CE, Long KJ, van Kooten TG, Meyle J (1996) Tissue Eng 2:241
116. Kono Y, Yang S, Roberts EA (1997) In Vitro Cell Dev Biol Anim 33:467
117. Moghe PV, Ezzell RM, Toner M, Tompkins RG, Yarmush ML (1997) Tissue Eng 3:1
118. Bock KW, Gschaidmeier H, Bock-Hennig BS, Eriksson LC (2000) Toxicology 144:51
119. Naruse K, Sakai Y, Nagashima I, Jiang GX, Suzuki M, Muto T (1996) Int J Artif Organs 19:605
120. Yuasa C, Tomita Y, Shono M, Ishimura K, Ichihara A (1993) J Cell Physiol 156:522
121. Wu FJ, Friend JR, Hsiao CC, Zilliox MJ, Ko W, Cerra FB, Hu W (1996) Biotechnol Bioeng 50:404
122. Torok E, Pollok JM, Ma PX, Kaufmann PM, Dandri M, Petersen J, Burda MR, Kluth D, Perner F, Rogiers X (2001) Cells Tissues Organs 169:34
123. Pelham RJ Jr, Wang Y (1997) Proc Natl Acad Sci USA 94:13661
124. Wang YL, Pelham RJ Jr (1998) Methods Enzymol 298:489
125. Beningo KA, Lo CM, Wang YL (2002) Methods Cell Biol 69:325
126. Weisz OA, Schnaar RL (1991) J Cell Biol 115:485
127. Weigel PH, Oka JA (1991) Biochem Biophys Res Commun 180:1304
128. Flanagan LA, Ju YE, Marg B, Osterfield M, Janmey PA (2002) Neuroreport 13:2411

129. Jamney PA, Yeung T, Flanagan LA (2001) ASME, Bioeng Div 50:709
130. Dealy JM, Wissbrun KF (1990) Melt rheology and its role in plastics processing. Van Nostrand Reinhold, New York
131. Bhadriraju K, Hansen LK (2000) Biomaterials 21:267
132. Tranquillo RT, Durrani MA, Moon AG (1992) Cytotechnology 10:225
133. Friedl P, Zanker KS, Brocker EB (1998) Microsc Res Tech 43:369
134. Wang N, Ostuni E, Whitesides GM, Ingber DE (2002) Cell Motil Cytoskeleton 52:97
135. Torok N, Urrutia R, Nakamura T, McNiven MA (1996) J Cell Physiol 167:422
136. Michalopoulos GK, Bowen W, Nussler AK, Becich MJ, Howard TA (1993) J Cell Physiol 156:443
137. Stolz DB, Michalopoulos GK (1997) J Cell Physiol 170:57
138. Nakamura T (1991) Prog Growth Factor Res 3:67
139. Michalopoulos GK, Bowen WC, Zajac VF, Beer-Stolz D, Watkins S, Kostrubsky V, Strom SC (1999) Hepatology 29:90
140. Fukujin H, Fujita T, Mine T (2000) Biochem Biophys Res Commun 278:698
141. Boylan JM, Gruppuso PA (1998) J Biol Chem 273:3784
142. Semler EJ, Moghe PV (2001) Biotechnol Bioeng 75:510
143. Dunn JC, Yarmush ML, Koebe HG, Tompkins RG (1989) Faseb J 3:174
144. Semler EJ, Lancin P, Dasgupta A, Moghe PV (2005) Biotechnol Bioeng 89:296

Adv Biochem Engin/Biotechnol (2006) 102: 47–90
DOI 10.1007/b137240
© Springer-Verlag Berlin Heidelberg 2005
Published online: 25 October 2005

Polymers as Biomaterials
for Tissue Engineering and Controlled Drug Delivery

Lakshmi S. Nair[1] · Cato T. Laurencin[2,3,4] (✉)

[1]Department of Orthopaedic Surgery, College of Medicine, University of Virginia,
Charlottesville, VA 22903, USA

[2]Department of Biomedical Engineering, University of Virginia,
Charlottesville, VA 22908, USA
ctl3f@virginia.edu

[3]Department of Chemical Engineering, University of Virginia, Charlottesville, VA 22904,
USA

[4]The University of Virginia, 400 Ray C. Hunt Drive, Suite 330, Charlottesville, VA 22903,
USA

Abstract The advent of biodegradable polymers has significantly influenced the development and rapid growth of various technologies in modern medicine. Biodegradable polymers are mainly used where the transient existence of materials is required and they find applications as sutures, scaffolds for tissue regeneration, tissue adhesives, hemostats, and transient barriers for tissue adhesion, as well as drug delivery systems. Each of these applications demands materials with unique physical, chemical, biological, and biomechanical properties to provide efficient therapy. Consequently, a wide range of degradable polymers, both natural and synthetic, have been investigated for these applications. Furthermore, recent advances in molecular and cellular biology, coupled with the development of novel biotechnological drugs, necessitate the modification of existing polymers or synthesis of novel polymers for specific applications. This review highlights various biodegradable polymeric materials currently investigated for use in two key medical applications: drug delivery and tissue engineering.

Keywords Biodegradable polymers · Drug delivery · Polymeric biomaterials · Tissue engineering

1
Introduction

Biomaterials are materials intended to interface with biological systems to evaluate, treat, augment, or replace any tissue, organ, or function of the body [1]. The essential prerequisite to qualify a material as a biomaterial is that it should be biocompatible. Biocompatibility is the ability of a material to perform with an appropriate host response in a specific application [1]. The criteria for determining the biocompatibility of a material depend on its end-use application. Consequently, a wide range of materials encompassing all the classical materials such as metals, ceramics, glasses, and polymers have been investigated as biomaterials. Among these, polymers form a versatile class of biomaterials that have been extensively investigated for medical and related applications. This can be attributed to the inherent flexibility in synthesizing or modifying polymers matching the physical and mechanical properties of various tissues or organs of the body.

The development of polymeric biomaterials can be considered as an evolutionary process. Reports on the applications of natural polymers as biomaterials date back thousands of years [2]. However, the application of synthetic polymers to medicine is more or less a recent phenomenon. The use of polymeric biomaterials as we know them today started in the 1940s during the second world war [3]. One of the first attempts was the use of the biostable synthetic polymer poly(methyl methacrylate) (PMMA) as an artificial corneal substitute. Encouraged by initial successes, surgeons started using a variety of polymers for different applications such as blood contacting devices, hip joint replacements, and as intraocular lenses [3]. However, in most of these cases, physicians were limited to using off-the-shelf materials initially developed for

Table 1 Different polymeric materials and their biomedical applications

Polymer	Application
Poly(methyl methacrylate)	Intraocular lens, bone cement, dentures
Poly(ethylene terephthalate)	Vascular graft
Poly(dimethylsiloxane)	Breast prostheses
Poly(tetrafluoroethylene)	Vascular graft, facial prostheses
Polyethylene	Hip joint replacement
Polyurethane	Facial prostheses, blood/device interfaces

other applications. Table 1 shows some of the synthetic polymeric materials employed and their biomedical applications. Even though the application of these polymeric materials significantly improved the advancement of modern health care, the long-term biocompatibility of many of these materials remained a serious concern.

During the latter half of the twentieth century, material scientists began attempts to engineer novel polymeric materials or modify existing polymers which could exhibit biocompatibility and adequate mechanical properties suitable for specific biomedical applications. In addition, recent advances in biotechnology and pharmaceutical science opened novel frontiers in biomedical fields that demanded materials with bioactivity, biocompatibility, and in many cases transient existence. The transient existence of materials is highly preferred for in vivo applications such as implantable drug delivery systems and conduits for guiding or remodeling damaged tissues. The biostable polymers are less than optimal for these applications, as in most cases a second surgical procedure must be performed to remove the implant to overcome long-term biocompatibility issues. This led to the quest for novel biodegradable polymers as candidates for various medical applications that require transient existence of the material.

The objective of this review is to highlight the various biodegradable polymers currently used or under investigation for two major medical applications: tissue engineering and drug delivery.

2
Biodegradable Polymers

The shift from biostable polymers to biodegradable polymers for applications that require transient existence of materials in the human body can be considered as a quantum leap in biomaterial science. Biodegradable polymers are those which degrade in vitro and in vivo either into products that are normal metabolites of the body or into products that can be completely eliminated from the body with or without further metabolic transformations. The basic criteria of selection of a biodegradable polymer as a biomaterial are that its degradation products should be nontoxic, and that the rate of degradation and mechanical properties of the material should match the intended application.

As is evident, the advantages of biodegradable polymers compared to biostable polymers are that once implanted they obviate the need for a second surgical procedure as well as eliminate the long-term biocompatibility concern. In addition to this, the biodegradation may offer other advantages in many short-term medical applications. Thus, in orthopaedic applications mechanically incompatible implants such as metallic implants can sometimes lead to stress shielding, whereas biodegradable implants may slowly transfer the load as it degrades. Similarly in drug delivery systems, fine-tuning of drug release kinetics is possible by varying the degradation rate of the matrix polymer.

Biodegradable polymers can be broadly classified into natural and synthetic, based on their origin. Natural polymers seem to be the obvious choice for biomedical applications due to their excellent biocompatibility, as structurally they closely mimic native cellular environments, have unique mechanical properties, and are biodegradable by an enzymatic or hydrolytic mechanism. However, natural polymers have not been fully exploited in the biomedical field due to the inherent disadvantages associated with some of them such as risk of viral infection, antigenicity, unstable material supply, and batch-to-batch variation in properties [2]. Synthetic polymers, on the other hand, offer tremendous advantages over natural polymers from the material side. Due to their synthetic flexibility it is possible to develop polymers having a wide spectrum of properties with excellent reproducibility. Furthermore, fine-controlling of the degradation rate of these polymers is highly feasible by varying their structure.

2.1
Synthetic Biodegradable Polymers

This section discusses the various synthetic biodegradable polymers currently being investigated as drug delivery systems or as scaffolds for tissue engineering. The section highlights the synthetic route, mode of degradation, and applications of the polymers.

2.1.1
Aliphatic Polyesters

Aliphatic polyesters can be considered as representatives of synthetic biodegradable polymers. Synthesis of aliphatic polyesters by polycondensation of diols and dicarboxylic acids was reported as early as 1930 [4]. However, the low melting points, low hydrolytic stability, and low molecular weights of the polymers initially obtained severely limited their application. At the same time, the high hydrolytic instability of these polymers resulted in a multitude of applications for this polymer class in the biomedical field starting with absorbable sutures in the 1960s [5]. This new avenue revitalized interest in this class of polymers, and novel synthetic methods, as well as catalytic systems, were developed to obtain high molecular weight polymers with narrow molecular weight distributions [6–10]. Polyesters are now synthesized by the polycondensation of diacids and diols, self-polycondensation of hydroxyacids, or by the ring-opening polymerization of cyclic diesters, lactones, glycolides, and lactides. The commonly used monomers for aliphatic polyester synthesis for biomedical applications are lactide, glycolide, and caprolactone [11]. Poly(glycolic acid) (PGA) (Fig. 1a) was one of the initially investigated biodegradable polyesters for biomedical applications. It is a highly crystalline polymer with a melting point greater than 200 °C and a glass transition temperature (T_g) around 35–40 °C. Due to its high crystallinity, PGA shows high tensile strength and modulus but very low solubility in common organic solvents. The initial applications of PGA were directed toward developing biodegradable sutures, and the first biodegradable synthetic suture DEXON was developed in 1970 [12]. PGA was also investigated as a material for the development of internal bone fixation devices, and has been commercialized under the trade name Biofix. However, the high degradation rate of the polymer and low solubility coupled with the accumulation of acidic degradation products, which can lead to inflammatory reactions, limits its application in the biomedical field.

As the lactide is a chiral molecule, polymeric lactide exists as three isomers: L-lactide, D-lactide, and *meso*-lactide. Therefore, four different types of poly(lactic acid) (PLA) are available: poly(L-lactic acid), poly(D-lactic acid), poly(DL-lactic acid), which is obtained from the racemic mixture of L- and D-lactic acid, and the *meso*-poly(lactic acid). Among these only poly(L-lactic acid) and poly(DL-lactic acid) have been extensively investigated as biomaterials. Poly(L-lactic acid) (PLLA) (Fig. 1b) is a semicrystalline polymer where the degree of crystallinity depends on the molecular weight and processing parameters, and it exhibits high modulus and strength. The melting point of PLLA is around 170 °C and it has a T_g of about 60–65 °C. Poly(DL-lactic acid) (PDLLA), on the other hand, is an amorphous polymer due to the random distribution of the two isomeric forms along the polymer chain and has a T_g of about 55–60 °C. Due to its low degradation rate, better processability, and mechanical properties, PLA has been extensively investigated as a bone fixative and is commercially available under different trade names such as FIXSORB [13].

Poly(caprolactone) (PCL; Fig. 1c) is a semicrystalline polyester, with a melting temperature around 55–60 °C and a T_g of – 60 °C. PCL is of great interest as can be obtained from cheap starting material (caprolactone), has high solubility in organic solvents, low melting point and T_g and an

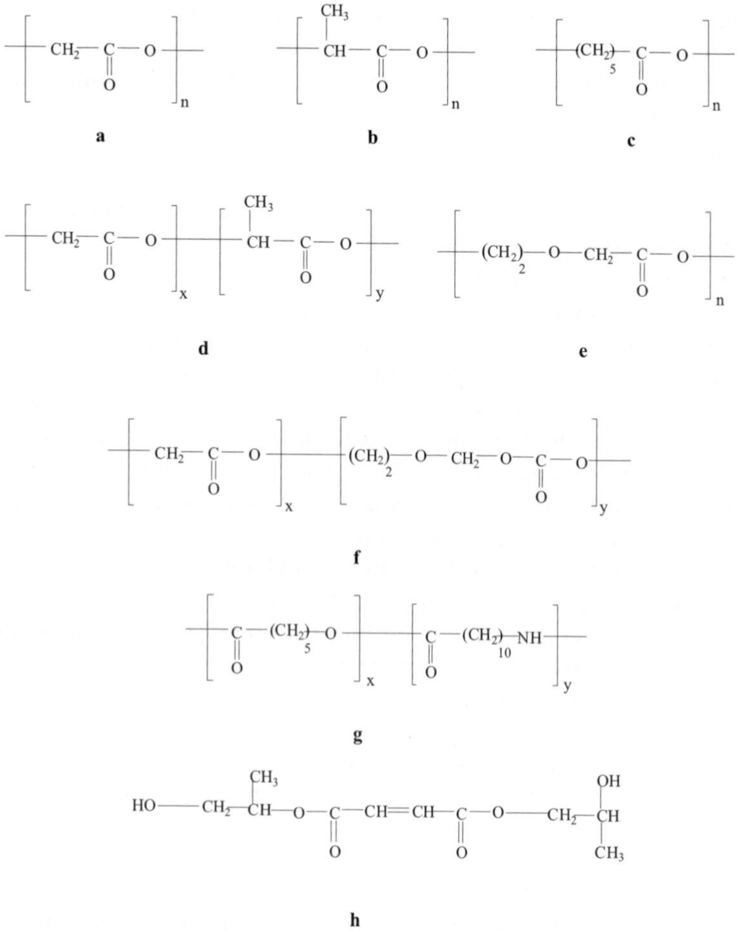

Fig. 1 Structure of aliphatic polyesters: **a** poly(glycolic acid), **b** poly(lactic acid), **c** poly(caprolactone), **d** poly(lactide-*co*-glycolide), **e** polydioxanone, **f** polyglyconate, **g** BAK; **h** polypropylene fumarate

exceptional ability to form blends with a variety of polymers. Due to the low degradation rate of PCL, it has been investigated as a material for the long-term controlled delivery of drugs. A long-term contraceptive device, Capronor, that could yield a zero-order release of levonorgestrel for over one year based on PCL has already been placed on the market [13].

Further property manipulation of these homopolyesters can be achieved by copolymerizing the respective monomers [13–15]. Copolymers of PLA and PGA (PLAGA) (Fig. 1d) have been extensively investigated for various medical applications such as sutures, bone pins, stents, drug delivery devices, and scaffolds for tissue engineering. The mechanical and degradation properties of these polymers can be tailored depending on the copolymer ratios.

Several multifilament sutures based on copolymers of glycolides with lactides are commercially available, such as Vicryl, PANACRYL, and POLYSORB.

The aliphatic polyesters undergo bulk degradation, where material is lost from the entire polymer volume at the same time due to water penetrating the bulk. So the rate of degradation of these polymers depends on the extent of water accessibility to the matrix rather than the intrinsic rate of ester cleavage. The water accessibility to the matrix depends on several factors such as the hydrophobicity/hydrophilicity of the polymer, the crystallinity of the polymer, and the dimension of the sample [14].

Occasionally, the acidic degradation products of these aliphatic polyesters have been implicated in adverse tissue reactions in certain biomedical applications. In drug delivery applications, the drug release pattern of macromolecules such as proteins from these matrices shows a marked initial burst release followed by a slow continuous release [16]. Further, the inactivation of sensitive molecules like proteins by the acid degradation products of these polymers has been reported [17]. This has led to the development of block or multiblock copolymers of the monomeric lactides/glycolides/lactones with other monomers to form poly(ether esters), poly(ester carbonates), poly(ester amides), and poly(ester urethanes) [13–15].

The well-known example of a poly(ether ester) is polydioxanone (Fig. 1e), which has been marketed as monofilament sutures under the trade name PDS. It is a semicrystalline, nontoxic polymer prepared by the ring-opening polymerization of p-dioxanone [18]. The polymer has a melting point around 115 °C and a T_g around – 10 to 0 °C. Copolymers of dioxanone with lactic or glycolic acid have also been synthesized to modulate the rate of degradation of the polymer. Another class of poly(ether ester) block copolymers that are gaining interest in the biomedical field due to their hydrophilicity, biocompatibility, and nontoxicity are those having poly(ethylene glycol) (PEG) units [19]. The semicrystalline nature of the ester domain combined with the flexible elastic and hydrophilic nature of PEG imparts novel properties to these polymers. The diblock copolymers of PLAGA and PEG are synthesized by the ring-opening polymerization of DL-lactide and glycolide with different molecular weights of monomethoxy polyethylene glycol as an initiator [20]. These diblock copolymers are being investigated as drug delivery systems. Copolymers of PEG with the aromatic block poly(butylene terephthalate) (PBT, Polyactive) are currently being investigated for several biomedical applications such as bone replacements, skin substitutes, and drug delivery devices, as the physical properties of these polymers can be readily varied by changing the polymer composition [21, 22].

The poly(ester carbonates) are prepared by the ring-opening polymerization of lactides/glycolides/lactones with trimethyl carbonate (TMC) or its derivatives [23]. These polymers are being investigated as drug delivery matrices as well as scaffolds for tissue engineering. The most extensively investigated poly(ester carbonate) is polyglyconate, which is a copolymer of glycolide with TMC (Fig. 1f) [11]. This polymer has been marketed as monofilament sutures under the trade name MAXON.

Poly(ester amides) form a class of biodegradable polymers having very good mechanical and thermal properties due to their polar nature, as well as the hydrogen bonding ability of the amide bonds and the biodegradability imparted by the ester bonds [24]. The degradation of poly(ester amides) has been shown to take place by the hydrolytic cleavage of the ester bonds, leaving the amide segments more or less intact. Several poly(ester amides) have been synthesized so far by the reaction of lactones/diacids with diamines/lactams (Fig. 1g) [25]. Poly(ester amides) based on caprolactam or hexamethylenediamine were commercialized under the trade name BAK in the mid-1990s [26]. Recently, attempts were made to increase the degradation rate of these polymers by incorporating amino acid units in the polymers [27].

Segmented polyurethane elastomers obtained by the reaction of diols with diisocyanates have been extensively investigated as materials for artificial hearts, small-vessel prostheses, and tracheal prostheses due to their excellent mechanical properties, chemical versatility, biocompatibility, and biostability. However, poly(ester urethanes) obtained using polyester diols are found to be hydrolytically unstable [28]. Thus, aliphatic polyesters such as lactide/glycolide copolymers or polycaprolactones are used as soft segments and polypeptides are used as hard segments in biodegradable poly(ester urethanes). A biodegradable elastic poly(ester urethane) (Degrapol) has been recently developed as a highly porous scaffold for tissue engineering applications [29].

Several biomedical applications, such as filling irregularly shaped bone defects or soft tissue repairs, require materials that possess a liquid or putty-like consistency, which can set and be molded into a desired shape under physiological conditions. Several in situ polymerizable systems, which can degrade with time due to the presence of hydrolytically sensitive bonds, have been developed for these applications. A class of unsaturated degradable polyesters that has been extensively investigated for orthopaedic applications is polypropylene fumarate (PPF, Fig. 1h), which can cross-link with itself or in the presence of other cross-linking agents [30, 31]. Several photo cross-linkable systems based on PPF have been investigated where the mechanical properties and the degradation rate can be varied by changing the cross-linking density.

2.1.2
Poly(ortho esters)

For drug delivery applications, the stability of drug in the matrix is equally as important as the rate of release of drug from the matrix. The biodegradable polymer that absorbs water during degradation is a concern, particularly for hydrolytically sensitive drugs. Hence researchers began exploring highly hydrophobic polymers with hydrolytically labile chemical bonds, where the polymer degradation can be confined to the surface, thereby protecting the drug within the matrix. Moreover, traditional polyester-based drug delivery systems often had difficulty attaining zero-order drug release kinetics. The drug release from a surface-eroding polymer matrix takes place by the erosion of the polymer rather than by diffusion from the matrix. This can give rise to both a zero-order release of the drug from the matrix and enhanced stability of the drug in the matrix.

Attempts were therefore made to develop various biodegradable hydrophobic polymer systems specifically for controlled drug delivery applications. Thus, poly(ortho esters) were developed by Alza Corporation as a hydrophobic polymer with a hydrolytically sensitive backbone, and marketed under the trade name Alzamer [32]. The high hydrophobicity of the matrix reduces the water penetration into the bulk, thereby confining the hydrolytic degradation to the surface. Thus, these polymers undergo surface erosion unlike the polyesters discussed above, and the rate of degradation can be controlled by using diols having different degrees of chain flexibility as well as by the incorporation of acidic and basic excipients. Four different families of poly(ortho esters) have been synthesized so far [33]. Poly(ortho ester) I is synthesized by the transesterification between a diol and diethoxytetrahydrofuran (Fig. 2a). One of its hydrolysis products γ-hydroxybutyric acid has an autocatalytic effect on the further degradation of the polymer. Poly(ortho ester) II is synthesized to overcome the autocatalytic effect of poly(ortho ester) I, and its degradation products are neutral molecules. Poly(ortho ester) II is obtained by the reaction of diols with diketene acetal 3,9-bis(ethylidene-2,4,8,10-tetraoxaspiro[5]undecane) (Fig. 2b) [34]. The reaction takes place spontaneously in a solvent such as THF, in the presence of an acid catalyst. However, the polymer is highly hydrophobic and usually needs an acid excipient to exhibit an appreciable degradation rate under physiological conditions. Poly(ortho ester) III can be synthesized by the direct polymerization of a triol with an ortho ester (Fig. 2c). The polymers exhibit a wide range of physical properties depending on the nature of the triol selected. However, these compounds suffer from drawbacks such as difficulty in synthesis and reproducibility. Poly(ortho ester) IV (Fig. 2d) has been developed as a modification of poly(ortho ester) II that could allow an appreciable degradation rate without the addition of an acid excipient [35]. This has been achieved by incorporating short segments based on lactic or glycolic acid on the polymer backbone, which can act as latent acids upon hydrolysis of the ortho ester linkage. Further, the rate of degradation of these polymers can be finely tuned from a few days to several months by varying the amount of the acid segment in the polymer backbone. Clinical trials are now in progress using this class of polymer, and some of the projected drug delivery applications of the polymer are in the treatment of postsurgical pain and ophthalmic diseases.

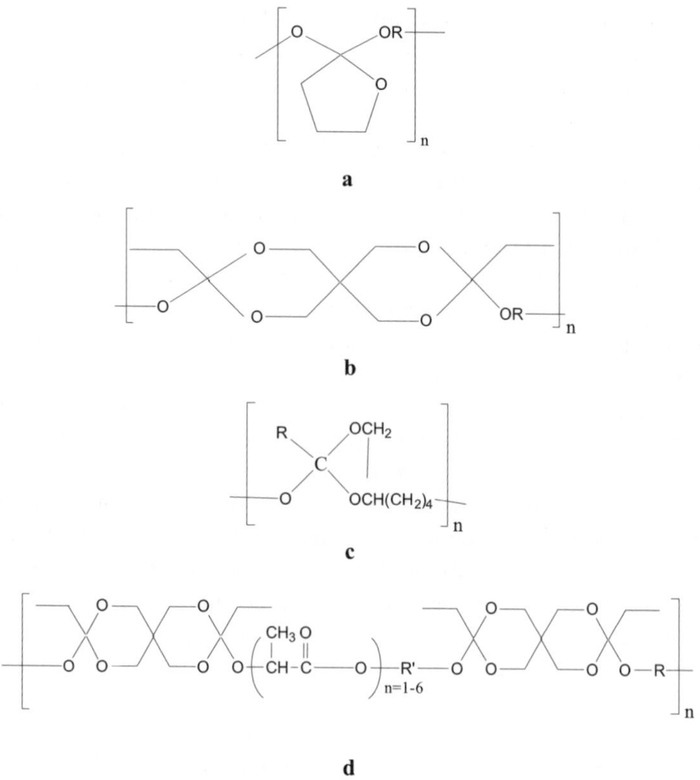

Fig. 2 Structure of poly(ortho esters): **a** poly(ortho ester I); **b** poly(ortho ester) II; **c** poly(ortho ester) III; **d** poly(ortho ester) IV

2.1.3
Polyanhydrides

Polyanhydrides form another class of surface-eroding polymers that have been extensively studied solely for biomedical applications. They form the most hydrolytically unstable polymers synthesized so far. The first synthesis of polyanhydrides was reported as early as 1909 by Bucher and Slade [36]. Hill and Carothers developed aliphatic polyesters and developed fibers of these materials for textile applications [37]. This was followed by the synthesis of aromatic polyanhydrides with very high hydrolytic stability by Conix [38]. The low hydrolytic stability of aliphatic anhydrides coupled with the low molecular weight of many of these polymers barred them from any industrial application until in 1980 Langer et al. proposed them as ideal candidates for drug delivery applications [39].

Polyanhydrides are synthesized via melt condensation of diacids/diacid esters, ring-opening polymerization of anhydrides, interfacial condensation, dehydrochlorination of diacids and diacid chlorides, or by the reaction of diacyl chlorides with coupling agents such as phosgene or diphosgene [40]. A variety of polymerization catalysts were recently developed which enables one to synthesize very high molecular weight polymers [40].

Like poly(ortho esters), polyanhydrides have a highly hydrophobic backbone and at the same time have a highly hydrolytically sensitive anhydride bond. This matrix hydrophobicity precludes water penetration into the matrix, and the degradation and erosion of the polymer are essentially confined to the surface as evidenced from the linear mass loss kinetics of the polymer during degradation [41]. This can be translated into zero-order drug release kinetics from many of these matrices when used as a drug delivery system.

Aliphatic homopolymers such as poly(sebacic anhydride) (PSA) are highly crystalline and have very high degradation rates that severely limit their application. Attempts were made to decrease the degradation rate of PSA by copolymerizing sebacic acid (SA) with highly hydrophobic aliphatic monomers such as linear fatty acid dimers (FAD) [40]. Recently nonlinear fatty acid dimers were investigated to circumvent the long in vivo life span of linear FAD-based polymers [40, 41]. Aromatic polyanhydrides, on the other hand, have very high hydrolytic stability and very high melting points, which make them difficult to process. Moreover, it has been established that the anhydride bonds between aliphatic carboxylic acids degrade faster than those of aromatic carboxylic acids. Thus, the degradation rate of polyanhydrides is a function of the polymer structure as well as the type of monomers used. This led to the development of various aliphatic–aromatic polyanhydrides whose properties can be tailored by varying the ratios of the two monomers [40, 41]. The most extensively investigated polyanhydride of this class is poly[(carboxy phenoxy)propane-sebacic acid] (PCPP-SA) (Fig. 3a). The high biocompatibility of these polymers has been established by Laurencin et al. [42]. Recently, the FDA has approved PCPP-SA as a delivery matrix for the controlled delivery of the chemotherapeutic agent BCNU to treat brain cancer under the trade name Gliadel.

Due to their rapid degradation and poor mechanical properties, the biomedical applications of polyanhydrides are currently confined to drug delivery devices. In order to improve the mechanical properties of polyanhydrides, imide segments have been incorporated to develop poly(anhydride-co-imide)s [43]. Several poly(anhydride-co-imide)s such as poly[pyromellitylimidoalanine-co-1,6-bis(p-carboxyphenoxy) hexane] (PMA ala:CPH) (Fig. 3b) have been investigated as scaffolds for tissue engineering, and the biocompatibility of these polymers was established in vitro and in vivo [44, 45]. Due to the availability of a wide range of diacid monomers, different types of polyanhydrides such as anhydrides containing ether, ester, and urethane linkages have been synthesized and investigated for various medical applications [46].

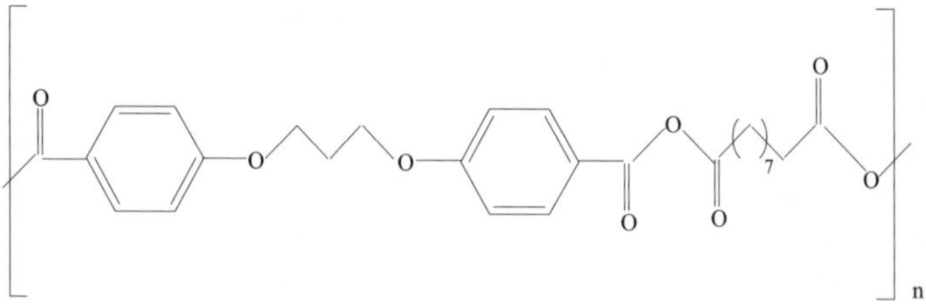

a

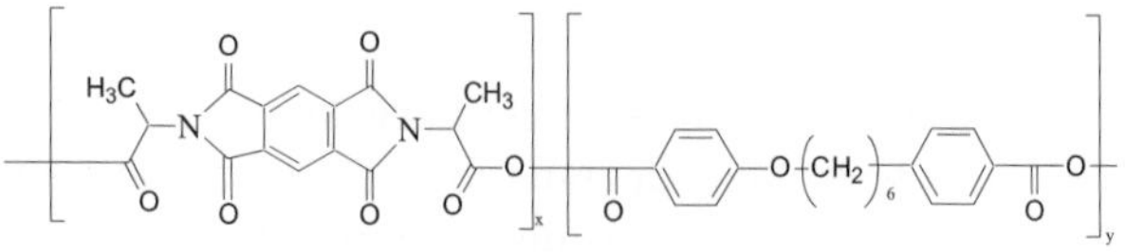

b

Fig. 3 Structure of polyanhydrides: **a** poly[(carboxy phenoxy)propane-sebacic acid]; **b** poly[pyromellitylimidoalanine-*co*-1,6-bis(*p*-carboxy phenoxy)hexane]

2.1.4
Poly(alkyl cyanoacrylates)

Poly(alkyl cyanoacrylates) form a unique class of biodegradable polymers where the carbon–carbon bonds in the polymer are cleaved by hydrolysis. The hydrolytic instability of carbon–carbon bonds in poly(cyano acrylates) can be attributed to the high inductive activation of methylene hydrogen atoms by the electron-withdrawing neighboring groups. These polymers exhibit properties that make them good candidates for drug delivery and drug targeting. In addition, they find applications as tissue adhesives, embolization agents, and haemostatic sealants. Poly(alkyl cyanoacrylates) (Fig. 4) are obtained by the anionic polymerization of alkyl cyanoacrylic monomers initiated by traces of moisture. Unlike other biodegradable polymers, poly(alkyl cyanoacrylates) exhibit high rates of degradation ranging from hours to days depending on the alkyl chain length of the polymer [47]. The lower alkyl derivatives such as poly(methyl cyanoacrylate), which degrades within a few hours, have been found to be toxic due to the formation of cyanoacetic acid and formaldehyde as degradation products. Hence, current investigations are being performed on higher alkyl derivatives such as butyl, octyl, or isobutyl cyanoacrylates. Nanospheres of these polymers, particularly butyl cyanoacrylates, have been extensively investigated for peptide and anti-cancer agent delivery [48]. Similarly, 2-octyl cyanoacrylate has been approved by the FDA as a tissue adhesive (Dermabond) for topical skin application [49].

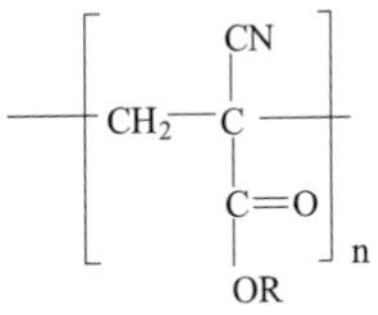

Fig. 4 Structure of poly(alkyl cyanoacrylate)

2.1.5
Poly(amino acids)

Synthetic poly(amino acids) have been investigated for various biomedical applications due to their structural similarity with naturally occurring proteins. Several homo- and copoly(amino acids) were synthesized and evaluated [50]. However, the high crystallinity, low degradation rate, unfavorable mechanical properties, and immunogenicity severely limited their biomedical applications [51]. Therefore, several different amino acid derived polymers have been developed as an attempt to overcome the unfavorable physicochemical and biological properties of synthetic amino acids. The amino acid derived polymers were synthesized by grafting amino acids on synthetic polymers, copolymerizing amino acids with other monomers, synthesizing block copolymers having amino acid sequences and poly(ethylene glycol), and by developing pseudo poly(amino acids) where the amino acid monomers are linked by non-amide bonds such as ester, carbonate, or iminocarbonate bonds [51, 52]. The most extensively studied pseudo amino acid polymer is the tyrosine-derived pseudo poly(amino acid) [51]. Tyrosine is a naturally occurring aromatic amino acid that can give good mechanical properties to the corresponding polymers. Tyrosine-derived polycarbonates are synthesized by polymerizing tyrosine in the presence of phosgene or bis(chloromethyl) carbonate triphosgene (Fig. 5). The physicochemical properties of these polymers can be tailored by varying the pendant alkyl ester chain. These are hydrophobic amorphous polymers with T_g less than 100 °C and decomposition temperatures around 300 °C [53]. The tyrosine-derived

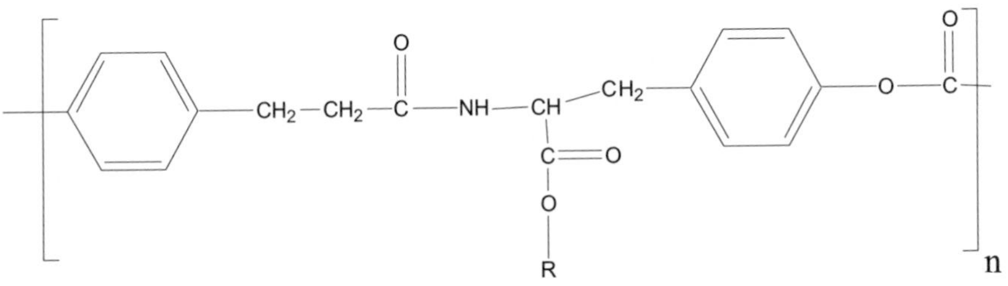

Fig. 5 Structure of tyrosine-derived polycarbonate

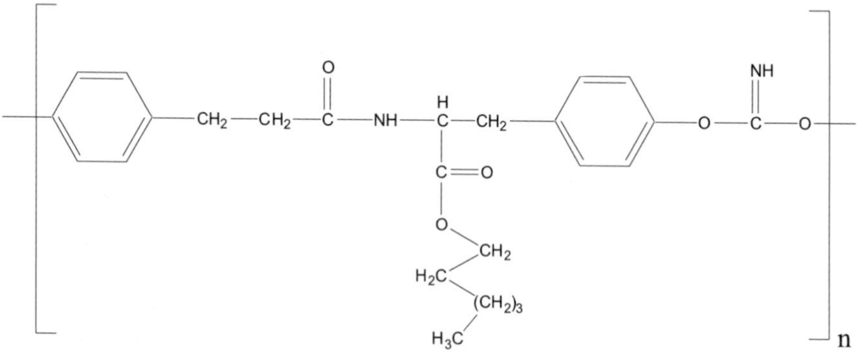

Fig. 6 Structure of tyrosine-derived poly(imino carbonate)

polycarbonates undergo hydrolysis at the carbonate bonds as well as the ester bonds in the pendant chain. The final degradation products of the polymer are tyrosine and the diols used to esterify the side chain. The in vivo degradation and excellent tissue compatibility of these polymers have been established in a rabbit transcortical model [54]. Due to the slow degradation rate of these polymers the tyrosine carbonates (desaminotyrosyl-tyrosine alkyl esters, DTH) have been investigated as matrices for the long-term delivery of drugs such as dopamine, as the polymer has structural similarity to the drug. The unique osteocompatibility of poly(DTH-carbonates) has been confirmed in a canine model [55]. This particular polymer can find wide applications as an orthopaedic implant material and is currently under review by the US Food and Drug Administration [55].

Poly(imino carbonates) based on natural amino acid L-tyrosine (Fig. 6) or its derivatives have been studied as amorphous, biodegradable polymers having a high degree of mechanical strength and stiffness [56]. Among the imino carbonates studied, poly(DTH-imino carbonate) has been found to be a promising candidate for transient medical applications due to its better degradability, low processing temperature, high mechanical strength, and better processability.

2.1.6
Polyphosphazenes

All of the synthetic biodegradable polymers discussed above can be considered as organic polymers; however, several inorganic polymers where the polymer backbone contains atoms other than carbon, nitrogen, or oxygen are assuming great importance as biomaterials due to their unique properties. Polyphosphazenes form a versatile class of inorganic polymers due to their synthetic flexibility and high adaptability for applications. These are linear, high molecular weight polymers with an inorganic backbone of alternating phosphorus and nitrogen atoms, with each phosphorus atom bearing two organic side groups. The first synthesis of hydrolytically stable high molecular weight polyphosphazene was reported by Allcock and Kugel in 1965 [57]. This was achieved by the controlled thermal ring-opening polymerization of hexachlorocyclotriphosphazene in the melt to form polydichlorophosphazene, followed by the substitution of the chlorine atoms of the polymer by organic or organometallic nucleophiles (Fig. 7, where R = alkoxy, aryloxy, or amino groups). The unique feature of polyphosphazenes is that the side groups play a crucial role in determining their physical properties [58]. The availability of a wide range of organic nucleophiles offers the opportunity to create a large number of polymers from polydichlorophosphazene. In addition, the macromolecular substitution route allows for the introduction of two or more different side groups on polydichlorophosphazene by simultaneous or sequential substitution, which results in polymers exhibiting a wide spectrum of properties.

Most of the initially developed polyphosphazenes were hydrolytically stable. However, the phosphorus–nitrogen backbone can be rendered hydrolytically unstable when substituted with appropriate organic side groups on the phosphorus atoms [59]. Thus, organic side groups such as amino acid esters, amines, imidazole, or alkoxide groups impart degradability to polyphosphazenes. The degradation products of these polymers were found to be neutral and nontoxic, with the degradation products being phosphates, ammonia, and the corresponding side groups. The degradation rate of the polymers can be fine-tuned by incorporating fewer or more highly hydrolytically sen-

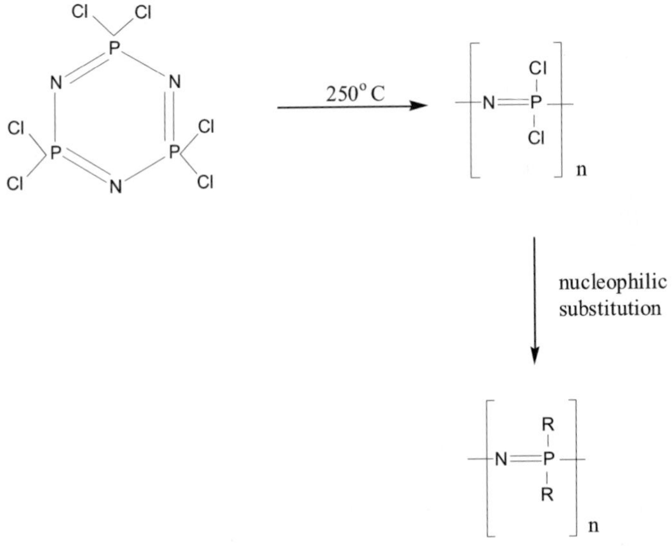

Fig. 7 Scheme showing the synthesis of polyphosphazene

sitive groups [60]. The degradation rate of these polymers depends on the nature and ratio of the side groups, pH of the surrounding medium, temperature, and solubility of the degradation products. The biocompatibility of many biodegradable polyphosphazenes have been established by in vitro and in vivo methods [60–64].

Most of the biodegradable polyphosphazenes are soluble in common organic solvents, which makes processing of the polymer highly feasible. Biodegradable polyphosphazenes are now being extensively investigated as matrices for drug delivery applications, particularly protein delivery [65, 66]. The synthetic versatility of polyphosphazenes makes it possible to design novel polymers with a desired degradation profile, which can be translated into designing drug release systems based primarily on diffusion, erosion, or a mixture of erosion and diffusion. Further, the good biocompatibility of these polymers makes them preferred candidates for tissue engineering. Several biodegradable polyphosphazenes are now under investigation for bone and neural tissue engineering applications [60–63]. The pentavalency of phosphorus in polyphosphazenes, unlike organic biodegradable polymers, provides active sites to which drug molecules can be attached, thus enabling the development of prodrugs as well as targeted delivery systems. Several monolithic drug delivery matrices based on biodegradable polyphosphazenes are currently undergoing advanced clinical trials [63, 64].

2.1.7
Polyphosphoesters

Polyphosphoesters form another novel class of inorganic polymers with a unique backbone consisting of phosphorus atoms attached to either a carbon or an oxygen atom. This class of polymers was developed in the 1970s by Penczek and his colleagues [67]. These polymers can be considered as analogs of nucleic acids and teichoic acids and are synthesized by a variety of routes such as ring opening, polycondensation, and polyaddition. Figure 8 shows the general structure of the polyphosphoester where R and R' can be varied to develop polymers with a wide range of physicochemical properties. There are three different classes of polymers belonging to the polyphosphoester family: polyphosphites (Fig. 8, $R' = H$), polyphosphonates (Fig. 8, $R' = $ alkyl/aryl group), and polyphosphates (Fig. 8, $R' = $ aryloxy/alkoxy group). Further-

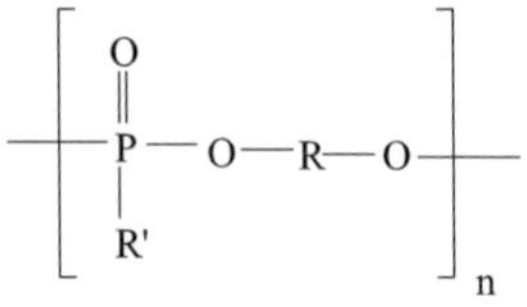

Fig. 8 Structure of polyphosphoesters

more, property manipulations of these polymers are possible by copolymerizing the phosphoesters with other monomers to form novel degradable polymers. Thus, copolymers of polyphosphoesters with DL-lactide [poly(lactide-co-ethyl phosphate)] and poly[bis(hydroxyethyl)terephthalate-ethyl orthophosphorylate/terephthaloyl chloride) (P[BHET-EOP/TC]) have been extensively investigated for biomedical applications [67, 68]. Polyphosphoesters degrade under physiological conditions due to the hydrolytic and enzymatic cleavage of the phosphate bonds in the backbone. The ultimate breakdown products of these polymers are phosphate, alcohol, and diols. The physicochemical properties of these polymers can be readily altered by varying either the backbone or side-chain organic components. The pentavalency of phosphorus atoms, as in the case of polyphosphazenes, allows for the chemical linkage of drugs or protein molecules to the polymer backbone, thereby enabling the development of novel polymer prodrugs. Most of these polymers show very high cytocompatibility in vitro and good tissue compatibility, as evidenced by low fibrous tissue encapsulation of the polymer in vivo [68]. Further, the matrices of poly(lactide-co-ethyl phosphate) showed near-zero-order release profiles of chemotherapeutic drugs such as paclitaxol, which makes them promising candidates for drug delivery applications, and they are now undergoing clinical trials under the trade name PACLIMER [69]. Polyphosphoesters are currently being investigated as nerve conduits and as matrices for drug and gene delivery applications [69].

2.2
Natural Biodegradable Polymers

This section deals with the various natural biodegradable polymers currently being investigated as matrices for controlled drug delivery or as scaffolds for tissue engineering. The section highlights the source, mode of degradation, and applications of the respective polymers.

2.2.1
Polysaccharides

Polysaccharides are high molecular weight polymers having one or more monosaccharide repeating units. Some of the advantages associated with polysaccharides are their wide availability, cost effectiveness, and wide range of properties and structures. Further, most of the polysaccharides can be easily modified due to the presence of reactive functional groups along the polymer chain. Their biodegradability, biocompatibility, and water solubility, combined with the ability to form hydrogels, make them excellent candidates for tissue engineering and drug delivery applications.

2.2.1.1
Cellulose

Cellulose forms the structural framework of plants and is isolated in the form of microfibrils. Cellulose is a linear polymer with repeating units consisting of D-glucose in $^4C_1$ conformation (Fig. 9). The cellulose can undergo enzymatic degradation resulting in the formation of D-glucose units. Even though it is a linear polymer, cellulose is insoluble in common solvents due to the presence of strong hydrogen bonding between polymer chains. However, the hydroxyl groups of cellulose are reactive and can be easily functionalized. Several derivatives of cellulose in the form of ethers, esters, and acetals, such as methyl cellulose, hydroxypropylcellulose, hydroxypropyl methyl cellulose, and carboxy methyl cellulose, have been investigated as candidates for various applications. All of these cellulose derivatives are soluble in a variety of solvents and can be easily processed into various forms such as membranes, sponges, and fibers. Cellulose membranes, due to their high diffusional permeability to most of the toxic metabolic solutes, have been extensively investigated as haemodialysis membranes [70]. Further, the good mechanical properties of cellulose coupled with the presence of reactive hydroxyl groups make cellulose an attractive matrix for fast protein purification [71].

Cellulose derivatives have been extensively investigated for biomedical applications as dressings in treating surgical incisions, burns, wounds, and various dermatological disorders. The dressings based on carboxymethylated cellulose fibers form a strong cohesive gel upon hydration and have been marketed under the trade name AQUACEL. Since the cellulose derivatives are water soluble and can form viscous jelly-like solutions, they have been investigated as ointment bases called "hydrogel bases" and have also been investigated as matrices for drug delivery applications [72].

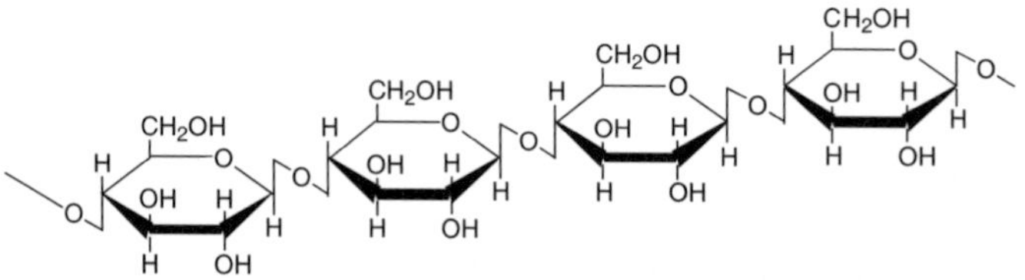

Fig. 9 Structure of cellulose

2.2.1.2
Starch

Starch is the carbohydrate reserve in plants and is usually isolated from corn, wheat, potato, tapioca, and rice. Structurally, starch is a combination of linear (amylose, 20–30%) and branched polymers (amylopectin, 70–80%) (Fig. 10).

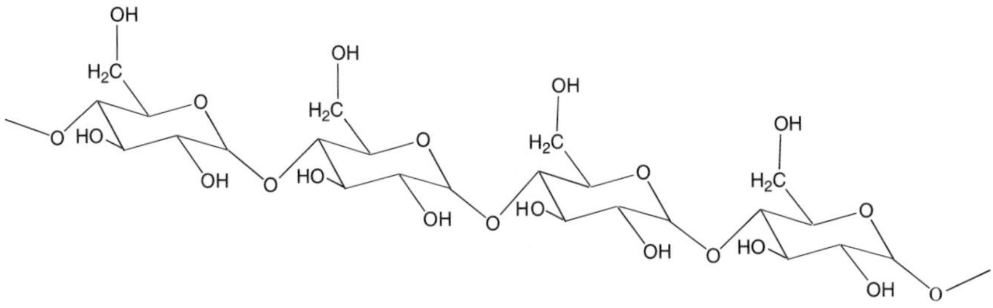

Fig. 10 Structure of starch

Both amylose and amylopectin consist of a single carbohydrate repeating unit of D-glucose. The linear amylose exhibits α (1 → 4) linkages with a relatively extended shape and amylopectin has α (1 → 6) linkages resulting in a compact, branched structure. The starch undergoes enzymatic degradation where the α (1 → 4) linkages are attacked by the enzyme amylase and the α (1 → 6) linkages are attacked by glucosidases. Starch can be easily processed after proper modification into thin films, fibers, or porous matrices suitable for various biomedical applications. Biodegradable starch-based blends have recently been suggested as potential candidates for biomedical applications, as they are biocompatible and biodegradable with degradation products comprising low molecular weight starch chains, fructose, and maltose [73]. A slow-release drug delivery system based on amylose-rich starch has been marketed under the trade name Contramid. Starch microspheres have been investigated as a bioadhesive drug delivery system for the nasal delivery of proteins [74]. Novel fabrication methods have been developed for making porous, starch-based scaffolds for tissue engineering [75]. Furthermore, the good cytocompatibility of starch-based matrices has been demonstrated in vitro [73].

2.2.1.3
Alginic Acid

Alginic acid is a linear heteropolysaccharide obtained from the cell walls of brown algae and is composed of D-mannuronic acid and L-guluronic acid (Fig. 11). It is commercially available as the sodium salt, sodium alginate. The polysaccharide contains block copolymer regions rich in mannuronic or guluronic acid as well as random copolymer regions of the two sugars. Due to the presence of the carboxyl groups along the polymer chain, alginates can form gels in the presence of divalent ions such as calcium ions. Gelation occurs by interaction of divalent cations with blocks of guluronic acid from different polysaccharide chains. The porous structure and high water absorbancy of calcium alginate gels combined with the haemostatic potential of alginates make them an attractive candidate for developing wound dress-

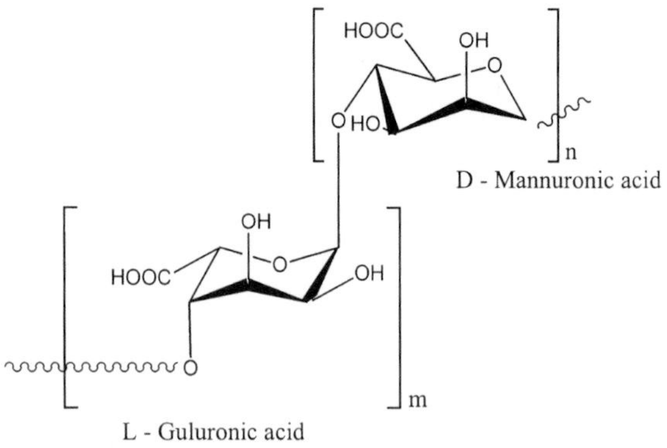

Fig. 11 Structure of alginic acid

ings [76]. Immobilization of cells in alginate is a well-established technology in a broad range of biotechnology and biomedical fields. Several alginate-based wound dressings are commercially available under trade names such as AlgiDERM, Algisite, Hyperion, and Kaltostat. Since the feasibility of making ionic gels of alginic acid is simple and rapid, alginate hydrogels have been extensively investigated as matrices for encapsulating cells for hybrid artificial organs as well as drug delivery systems [77]. Alginate beads have been extensively used for the controlled delivery of many cationic drugs and various growth factors. The drug loading and rate of release from these systems is greatly influenced by the ionic interaction between the drug and the alginate matrix. Polyelectrolyte complexes of alginate with other cationic polymers such as chitosan have also been extensively investigated for cell encapsulation or as drug delivery matrices [78]. Further, alginate hydrogels have been shown to exhibit very low interaction with cells or proteins. For that reason, alginates have been investigated as matrices for the encapsulation of chondrocytes, an anchorage-independent cell type. Chondrocytes encapsulated in alginate remain differentiated with spherical cell morphology and secrete the cartilage-specific markers collagen II and aggrecan [79].

2.2.1.4
Hyaluronic Acid

Hyaluronic acid is a naturally occurring linear anionic polysaccharide consisting of repeating disaccharide units. Hyaluronic acid forms an important component of articular cartilage and is widely distributed in the connective tissue as well as vitreous and synovial fluids of mammals. It is also abundantly present in the mesenchyme of developing embryos. Chemically hyaluronic acid consists of D-glucuronic acid and 2-acetamido-2-deoxy-D-glucose monosaccharide units (Fig. 12). The polymer is water soluble and forms very viscous solutions. Hyaluronic acid possesses several properties

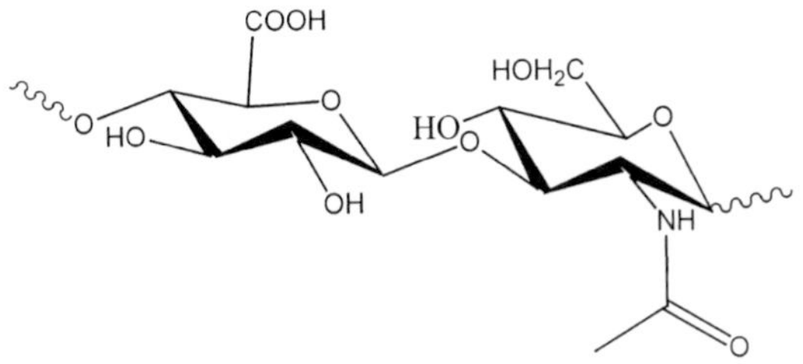

Fig. 12 Structure of hyaluronic acid

that make it an ideal candidate for wound dressing applications [80]. It can act as a scavenger for free radicals in wound sites, thereby modulating inflammation, it can interact with a variety of biomolecules, it is a bacteriostat, and it can be recognized by receptors on a variety of cells associated with tissue repair. Consequently, cross-linked hyaluronic acid gels or hyaluronic acid derivatives such as ethyl/benzyl (HYAFF) esters have been extensively investigated for wound dressing applications. The rate of degradation and the solubility of the hyaluronic acid can be effectively controlled by changing the extent of esterification [81]. Furthermore, these derivatives can be fabricated into a variety of shapes such as membranes, fibers, sponges, and microspheres. Hyaluronic acid is known to accelerate tissue repair by promoting mesenchymal and epithelial cell migration and differentiation, thereby enhancing collagen deposition and angiogenesis [80]. Due to these versatile properties, hyaluronic acid has been extensively used in a variety of clinical products. High molecular weight viscoelastic hyaluronic solutions (AMVISC and AMVISC PLUS) are used to protect delicate tissue in the eye during cataract extraction, corneal transplantation, and glaucoma surgery. It can act as a vitreous substitute during retina reattachment surgery. Injectable formulations of hyaluronic acid (SYNVISC, ORTHOVISC) have been developed to relieve pain and improve joint mobility in patients suffering from osteoarthritis of the knee. Another viscous formulation of hyaluronic acid containing fibroblast growth factor (OSSIGEL) has been developed to accelerate bone fracture healing and is currently undergoing advanced clinical trials.

2.2.1.5
Chitin and Chitosan

Chitin is a naturally occurring polysaccharide which forms the outer shell of crustaceans, insect exoskeletons, and fungal cell walls. It is the second most abundant natural polymer, cellulose being the first. It is a $(1 \rightarrow 4)$ β-linked glycan composed of 2-acetamido-2-deoxy-D-glucose (Fig. 13a). The N-acetyl glucosamino groups in chitin show structural similarity to hyaluronic acid which has very high wound healing potential. Chitin is also known to have accelerating effects on the wound healing process. Chitin fibers, mats,

Fig. 13 **a** Structure of chitin **b** Structure of chitosan

sponges, or membranes have been investigated as wound dressing materials. As chitin is insoluble in common solvents, its deacetylated derivative chitosan is the most extensively studied polymer for biomedical applications. Chitosan is a semicrystalline linear polymer of $(1 \rightarrow 4)$ β-linked D-glucosamine residues with some randomly distributed N-acetyl glucosamine groups (Fig. 13b). Chitosan is completely soluble in aqueous solutions with pH lower than 5.0 [82]. It undergoes biodegradation in vivo enzymatically by lysozyme to nontoxic products [82]. The rate of degradation of chitosan depends inversely on the degree of acetylation and crystallinity of the polymer [82]. The easy processability of chitosan coupled with its versatile properties makes it an attractive material for various medical applications. Chitosan has been extensively investigated as a wound and burn dressing material due to its easy applicability, oxygen permeability, water absorptivity, haemostatic property, and ability to induce interleukin-8 from fibroblasts, which is involved in the migration of fibroblasts and endothelial cells [83]. Incorporation of antibacterial agents into these wound dressings significantly improves the performance of chitosan-based dressings [84]. Chitosan can form polyelectrolyte complexes with various anionic polymers such as alginic acid/hyaluronic acid and, due to the presence of reactive amino and hydroxyl groups, it can be cross-linked using a variety of chemical cross-linking agents. Cross-linked chitosan hydrogel matrices form attractive materials for drug delivery applications where the rate of drug release can be controlled by varying the cross-linking density [82]. The feasibility of forming porous scaffolds may permit wide applications for this polymer in tissue engineering. This is particularly true for bone tissue engineering applications, as chitosan is known to support osteoblast proliferation and phenotype expression [85].

In addition to the polymers discussed above a variety of biodegradable polysaccharides, such as chondrotin sulfate, dextran, mannan, inulin, and agarose, have also been under investigation for various biomedical applications.

2.2.2
Proteins

Proteins are high molecular weight polymers having amino acid repeating units where the amino acids are joined together by characteristic peptide linkages (Fig. 14). Since the major components of soft and hard tissues in the human body are composed of proteins, these materials have been extensively investigated for various applications such as sutures, haemostatic agents, scaffolds for tissue engineering, and drug delivery.

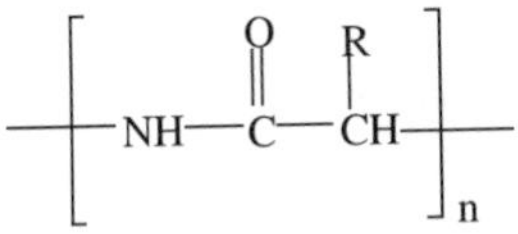

Fig. 14 Structure showing the peptide linkages in proteins

2.2.2.1
Collagen

Collagen forms the most abundant protein in the human body, being the major component of bone, skin, ligament, cartilage, and tendon. It also forms the structural framework for other tissues such as blood vessels. The basic unit of collagen is a polypeptide consisting of the repeating sequence of glycine, proline, and hydroxyproline. This polypeptide combines with 12 others to form the left-handed triple helix structure in collagen. The name collagen comprises a group of closely related proteins having similar structural characteristics. At least 22 different types of collagen have been identified so far in the human body. However, the most extensively investigated ones are collagen types I, II, III, and IV. Type I collagen is present in tendons, ligaments, and bones and consists of striated fibers between 80 and 160 nm in diameter. Type II collagen consists of fibers that are less than 80 nm in diameter and present in cartilage and intervertebral disks. Type III forms reticular fibers in tissues and strengthens the walls of hollow structures such as arteries, intestine, and uterus. Type IV is a highly specialized form of collagen present as a loose fibrillar network in the basement membrane [86].

Due to its enzymatic degradability and unique physicochemical, mechanical, and biological properties, collagen has been extensively investigated for various medical applications. Further, it can be processed into different forms such as sheets, tubes, sponges, powders, fleeces, injectable solutions, and dispersions. The use of collagen as a suture material dates back a millennium, and one form of it, catgut, is still in use for surgery [87]. Due to its biocompatibility, biodegradability, and ability to be cross-linked by a variety of agents, collagen is an attractive candidate for drug delivery applications. Collagen has been investigated in the form of collagen shields/particles in ophthalmology to deliver drugs to the eyes [88]. Collagen has also been used in ophthalmology as grafts for corneal replacement, suture material, bandage lenses, punctual plugs, or as viscous solutions for vitreous replacements during surgery.

The high thrombogenicity of collagen makes it a potential candidate as a haemostatic agent [89]. Recently, the FDA approved a haemostatic sealant consisting of bovine collagen and bovine thrombin for treatment of bleeding in cardiovascular and spinal surgical procedures (Sulzer-Spine Tech). Further, due to their versatile biological properties, collagen matrices can significantly improve cellular adhesion and proliferation. Consequently, collagen sponges have been extensively investigated as wound and burn dressings. An acellular collagen matrix obtained from human cadavers after chemical processing has been FDA approved for the treatment of burns as well as for reconstructive surgery, and has been marketed under the trade name Alloderm.

Collagen combined with proteoglycans is metabolically stable and has been extensively investigated as an artificial skin which accelerates wound healing [90]. A three-dimensional (3-D) porous matrix of cross-linked collagen and glycosaminoglycans (GAGs) has been FDA approved as a dermal matrix, and has been marketed under the trade name Integra. Other collagen-based skin equivalents approved by the FDA consist of a collagen layer seeded with fibroblasts and keratinocytes (Apligraf and Orcel). Collagen-based matrices also find wide applications in tissue engineering and have been extensively investigated in combination with growth factors [91].

Due to its fibrous nature, collagen can withstand tensile loads and hence has been investigated for applications that require structural integrity. Thus, composites of collagen with hydroxyapatites have been investigated as scaffolds for bone tissue engineering due to their excellent biocompatibility and because they closely mimic the composition of natural bone. Collagraft, a composite of bovine type I collagen and hydroxyapatite/tricalcium phosphate granules has been approved by the FDA as a synthetic bone-graft substitute.

In spite of the versatile properties of this natural polymer, only very few products based on collagen are going into clinical trials. This can be attributed to several shortcomings associated with this polymer. The immunogenicity of collagen, which in turn depends on the source and processing techniques, is still a concern. Other concerns include the high cost of pure collagen, variability in the physicochemical and degradation properties depending on source and processing technique, and the threat of the transmission of infectious diseases such as bovine spongiform encephalopathy, as cattle form the main source of collagen.

2.2.2.2
Gelatin

Gelatin is prepared by the thermal denaturation of collagen, isolated from animal skin and bones in the presence of dilute acid. Structurally, gelatin consists of 19 amino acids joined by peptide linkages, and the enzymatic degradation of the gelatin results in the formation of the corresponding amino acids. Biologically, gelatin does not show antigenicity and has high haemostatic properties. Further, the harsh acidic or basic conditions used to prepare gelatin eliminate many of the adverse properties associated with collagen described above. Due to its biosafety, gelatin has been in clinical usage for a considerable time and is being used in a variety of medical products regulated by the FDA.

Gelatin, like collagen, exhibits haemostatic properties and has been investigated for the development of biological glue composed of resorcinol and formaldehyde [92]. Commercialization of this glue has been limited by reports of cytotoxicity due to the release of formaldehyde upon degradation. A topical haemostatic agent based on gelatin and thrombin (FloSeal) was approved by the FDA in 1999. Gelatin can be easily cross-linked by a variety of cross-linking agents and forms a hydrogel capable of imbibing large quantities of water. Further, depending on the processing conditions, the electrical nature of gelatin can be varied resulting in gelatin with different isoelectric points, which allows the possibility of forming polyelectrolyte complexes with other ionic polymers. Because of the ease of processability, biodegradability, and hydrogel properties of chemically cross-linked or polyelectrolyte complexes of gelatin, they have been extensively investigated as drug delivery matrices. The high cytocompatibility of gelatin makes it a suitable candidate for tissue engineering applications, particularly as delivery matrices for growth factors [93, 94].

2.2.2.3
Albumin

Albumin is the protein of highest concentration in blood plasma. The primary function of albumin is to carry hydrophobic fatty acid molecules around the bloodstream. The polymer consists of a single polypeptide chain and exists mainly in the α-helical form. Albumin, like gelatin, can be easily processed into membranes, microspheres, or nanospheres due to its solubility as well as to the presence of reactive functional groups along the polymer chain. Since albumin inhibits fibrinogen adsorption and platelet aggregation, it has been extensively investigated as coatings for biomaterials to improve their blood compatibility. Due to its high blood compatibility albumin has been extensively investigated as a matrix for intravascular drug delivery systems [95]. Recently, a surgical tissue adhesive based on bovine albumin and glutaraldehyde has been approved by the FDA for reapproximating the layers of large vessels such as aorta, and femoral and carotid arteries (CryoLife Inc.).

2.2.3
Bacterial Polyesters and Polyamides

Bacterial polyesters and polyamides are polymers produced by microorganisms in response to particular nutrient and culture conditions. These are nontoxic, biocompatible, biodegradable materials and are being investigated for a variety of applications in medical and related fields.

2.2.3.1
Polyhydroxyalkanoates

Polyhydroxyalkanoates (PHA) are attracting great attention due to their structural diversity and close analogy to plastics [96]. These biodegradable polyesters are synthesized by many gram-positive and gram-negative bacteria. These polymers are stored intracellularly to about 90% of the cell weight under conditions of nutrient stress and act as a carbon and energy reserve for the cells [97]. Polyhydroxyalkanoates are commonly composed of β-hydroxy fatty acids (Fig. 15) where the R group changes from methyl to tridecyl. Poly(3-hydroxy butyrate) (PHB; Fig. 15, R = CH_3) is the most investigated PHA. In addition to homopolymers, copolymers such as those having hydroxyl butyrate and hydroxyl valerate (Fig. 15, R = CH_2CH_3) units could be formed by changing the composition of the nutrient medium. The PHAs are highly crystalline, thermoplastic polymers and are insoluble in water. They exhibit optical activity, isotacticity, and piezoelectric properties. Further, they are biodegradable, nontoxic, and elicit minimal inflammatory response in vivo. The copolymers degrade faster than the homopolymers. Under aerobic conditions the polymer degrades into carbon dioxide and water. The PHBs are soluble in common organic solvents and can be processed into membranes, fibers, or microspheres. Due to their biocompatibility, processability, and degradability these polymers have been investigated as matrices for drug delivery applications and tissue engineering [98]. The piezoelectric properties associated with these polymers make them attractive materials for orthopaedic applications, such as bone plates, as they may stimulate bone growth [96].

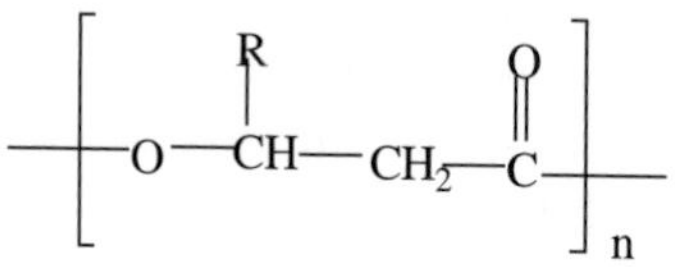

Fig. 15 Structure of polyhydroxyalkanoates

2.2.3.2
Poly(γ-glutamic Acid)

Poly(γ-glutamic acid) (γ-PGA) is an anionic, homopolyamide produced by various strains of *Bacillus*. The polymer is made of D- and L-glutamic acid units connected by amide linkages between α-amino acid and γ-carboxylic acid groups (Fig. 16). The favorable properties of this polymer such as water solubility, biodegradability, and nontoxic nature of degradation products make it an attractive candidate for medical applications such as drug delivery. Further, the availability of reactive carboxyl groups on the polymer chains allows the covalent immobilization of drugs. A novel drug delivery system based on covalent immobilization of the well-known anticancer drug Taxol on poly(glutamic acid) has been developed [99]. Preclinical evaluation showed promising results with the delivery system. Recently, a novel biological adhesive has been developed based on gelatin and poly(glutamic acid), which shows promise as a surgical adhesive and haemostatic agent [100, 101]. The water solubility of this polymer makes it a viable candidate for hydrogel preparation. The hydrogels of poly(glutamic acid) prepared by γ-irradiation are under investigation as drug delivery matrices [102].

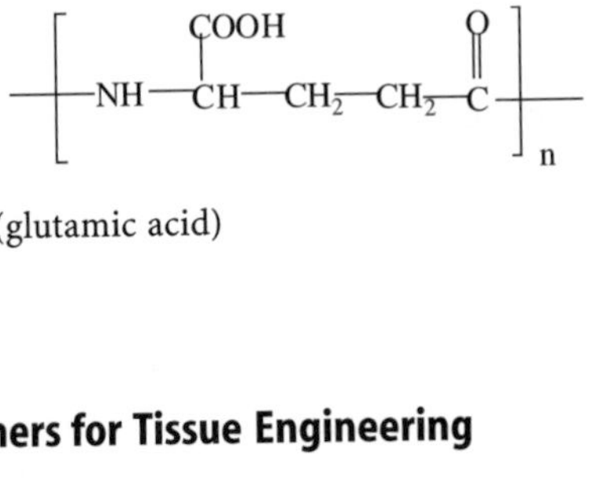

$$\left[-NH-\underset{\underset{COOH}{|}}{CH}-CH_2-CH_2-\overset{\overset{O}{||}}{C} \right]_n$$

Fig. 16 Structure of poly(glutamic acid)

3
Biodegradable Polymers for Tissue Engineering

3.1
Scope of Tissue Engineering

Tissue damage or loss due to congenital diseases, trauma, or accidents and end-stage organ failures are the two major causes of illness and death worldwide. The treatment modalities conventionally employed for these are the transplantation of tissues and organs (autograft, allograft/xenograft) or the use of mechanical assist devices. Although these approaches significantly improve patient survival, they suffer from serious limitations. Autografts (tissue isolated from the same patient) possess limitations such as donor-site morbidity as well as the associated risk of infection and limited availability. Allografts (tissue or organ isolated from another individual of the same species) and xenografts (tissue or organ isolated from another species) pose serious constraints due to immunoincompatibility, which necessitates the patients undergoing lifelong immunosuppression treatment. Further, they pose increased risk of infection, viral disease transmission, tumor development, and many associated side effects. Moreover, tissue or organ transplantation is highly expensive and complex surgery, and is further limited by donor shortages and the limited time the organs can be preserved outside the body. The success of mechanical replacement devices or total artificial organs is seriously limited by thromboembolization, associated infection, and durability [103].

Many of the shortcomings of organ or tissue transplantation can be logically overcome if it is possible to engineer and develop tissues in vitro that can specifically meet the needs of individual patients. Thus, tissue engineering has emerged as a novel therapeutic strategy to repair or reconstruct damaged tissues and organs. Tissue engineering has been defined as the application of biological, chemical, and engineering principles to the repair, restoration, or regeneration of living tissues using biomaterials, cells, and growth factors alone or in combination [104].

Natural tissues are three-dimensional (3-D) structures composed of cells surrounded by extracellular matrix (ECM). The ECM forms the supporting matrix for the cells to reside and the cell–cell and cell–ECM contacts play an important role in maintaining cell differentiation and function. The tissue engineering approach utilizes mainly a cell-matrix construct to develop functioning tissues.

Extensive research has been performed to develop materials that can be used as viable scaffolds for tissue engineering. Several parameters should be taken into consideration when choosing materials for scaffold fabrication depending on the intended application. The primary function of a scaffold is to provide structural support to growing cells and to provide a 3-D environment to guide the formation of new tissue. In order to provide structural support the material should have appropriate mechanical properties for the intended application and for the intended period of time. In most applications a tissue-engineered matrix should undergo degradation in vivo to be replaced by a new matrix synthesized by exogenously seeded cells or by migrating endogenous cells. Thus, biodegradable polymers are preferred candidates for developing tissue-engineered constructs. In any case, the rate of degradation of the matrix should not be higher than the matrix synthesis in vivo. Further, the matrix should have a 3-D architecture for the seeded cells to organize into tissues. Moreover, the volume of tissue that can be developed using a scaffold depends significantly on the pore size and pore tortuosity of the 3-D matrices, since cell infiltration and nutritional/gas exchange are crucial for tissue development in vitro and for angiogenesis in vivo. However, depending on the type of tissue to be developed, the optimal property requirements may vary. Therefore, extensive research has been performed in developing various scaffold fabrication techniques [105, 106]. Recently computer-aided design and computer-aided manufacturing (CAD/CAM) systems have been developed as a promising technique for creating complex scaffolds for tissue engineering. Here a computer model of the desired tissue is first developed, followed by a robotic manufacturing system which develops the scaffold based on the design [107, 108]. Apart from the mechanical properties and 3-D architecture of the matrix, the cell adhesiveness of the scaffold surface plays a crucial role in determining the organization of cells on the scaffold. The cell adhesiveness on matrices mainly depends on the physical and chemical properties of the scaffold surface as well as the properties of the cell type under investigation. Further, the matrix should also modulate adhered cells to differentiate and maintain phenotype expression. The above discussion clearly shows that no single biomaterial can provide the multitude of properties required to fabricate matrices for engineering various tissues and organs. Therefore, the material selection and fabrication technique should be chosen based on the requirements of the intended application.

3.2
Engineered Tissues

Tissue engineering has currently developed into an active area of research. Tissue engineers are presently engaged in attempts to engineer virtually all the tissues of the human body. This section highlights the tissue engineering approaches employed to regenerate some selected organs of the human body.

3.2.1
Skin Regeneration

Skin is the largest organ of the human body and functions to protect the body from the external environment by maintaining temperature and haemostasis, as well as by performing immune surveillance and sensory detection [109]. Skin consists of two main layers, an outer epidermis composed of stratified squamous epithelium and an inner dermis composed of dense connective tissue and fibroblasts. Significant skin loss due to injury or illness leading to damage of dermal or subdermal tissues cannot heal properly and can lead to serious consequences. Tissue-engineered skin refers to skin products made up from cells and extracellular matrix alone or in combination with growth factors [110]. A variety of materials have been investigated as matrices including autologous, allogenic, and xenogenic tissues or synthetic and natural polymeric materials for skin tissue engineering. Ideally a graft material should promote adequate wetting and draping, thus eliminating dead space, exhibit good adherence to the wound, prevent bacterial invasion, and control fluid loss. Further, the material should have adequate mechanical properties, promote endogenous cell infiltration, promote angiogenesis, and degrade with time. Three different approaches are currently in use to create artificial skin: to recreate the epidermis, to recreate the dermis, and to recreate both the dermis and epidermis using a bilayer graft. Thin layers of keratinocytes as such or cells cultured on polyurethanes (Epicel) and hyaluronic ester membranes (LASERSKIN) have been developed as epithelial replacements. Dermal substitutes make use of biodegradable polymers such as collagen–glycosaminoglycan matrix covered with a silastic membrane as a barrier layer (Integra) as well as PLAGA (Vicryl) matrices seeded with human fibroblast cells (Dermagraft). A full-thickness graft has been developed using bovine collagen I matrices seeded with fibroblasts and keratinocytes called human skin equivalents (Apligraf). Multicenter clinical trials have revealed the accelerated healing capacity of chronic nonhealing venous stasis ulcers by the tissue-engineered graft [111].

3.2.2
Nerve Regeneration

Many diseases of the nervous system originate from injury to neurons and their extracellular environment. Neural tissue once damaged may not heal by true regeneration. Neural tissue engineering aims to enhance nerve regeneration using scaffolds, growth factors, and neuronal support cells or genetically engineered cells [110]. Currently, peripheral nerve regeneration of transected nerves that traverse a significant distance is performed by reconnecting the proximal and distal ends of injured nerves using autografts or tubular polymeric nerve guides. The tubular nerve guide provides directional guidance and acts as a barrier between the growing axons and surrounding environment. Initially nerve guidance channels made of nondegradable silicon tubes (SE) were commonly used for nerve regeneration. Biodegradable natural polymers such as laminin, collagen, and chondrotin sulfate, and synthetic polymers such as PLLA, PLAGA, PCL, and polyester urethanes are now replacing SE as they obviate the need for second surgery. Biodegradable polyphosphazenes have recently gained interest as materials for neural tissue engineering [112]. Preliminary studies showed significantly improved nerve regeneration, minimal scar tissue formation, and low foreign body reaction to these matrices compared to silicon implants. The performance of these nerve-guide channels can be further improved by preseeding them with neuronal cells (Schwann cells), precoating with neurotrophic factor,s or prefilling with synthetic biodegradable polymeric gels [110]. Prefilling nerve-guide tubes with natural polymers such as fibrin, collagen, or collagen–GAG mixtures significantly improves nerve growth [113]. Another attempt that shows considerable promise in nerve regeneration is prefilling nerve-guide tubes with hydrogels of natural polymers such as agarose or alginates functionalized with ECM protein analogs or oligopeptides [114]. Further, the release of growth factors from biodegradable nerve-guide channels can significantly improve nerve growth over greater distances [115].

3.2.3
Liver

The liver is the largest internal organ and its functions are to extract and metabolize toxic substances from the blood. The liver is a complex organ with several tissue types and hepatocytes. The most successful treatment for liver failure to date is orthotropic liver transplantation; however, it faces serious drawbacks such as donor shortages and rapid progression of disease [116]. Attempts are now under way to achieve liver replacement using isolated hepatocytes. Several culturing techniques for hepatocytes, such as culturing in suspensions and encapsulating them in polymeric microcapsules, have been investigated [117]. Several biodegradable polymers were investigated as substrates for hepatocyte culture in vitro. Encapsulation of hepatocytes in a polymer matrix provides an environment which would allow hepatocytes to remain differentiated. Collagen, which is the major component of ECM, has been extensively investigated for hepatocyte culture [118]. Calcium cross-linked alginate gels are another substrate extensively investigated for hepatocyte encapsulation [119]. Another natural polymer investigated for liver regeneration is chitosan [120]. Synthetic biodegradable polymers investigated for hepatocyte culture include PGA, PLA, and PLAGA [117]. Hybrid biomaterials formed from synthetic and natural biodegradable polymers in the form of highly porous sponges or fiber mesh have also been investigated for hepatocyte culture and transplantation [121, 122].

3.2.4
Cartilage

Articular cartilage is a hydrated biological soft tissue that lines the bony ends of all diarthroidal joints [79]. The primary function of cartilage is to distribute stresses in load-bearing sites and reduce friction during joint motion. Once injured, articular cartilage has limited capacity for repair due to the lack of blood supply in the tissue and the absence of undifferentiated cells for repair [79]. Attempts are currently under way to promote cartilage repair using the principles of tissue engineering. One cell-based repair procedure currently in clinical use is Carticel (Genzyme) which employs autologous chondrocytes. Several other approaches are currently being explored which make use of a combination of cells, biodegradable matrices, and growth factors for effective cartilage repair. The most commonly investigated synthetic biodegradable polymers as a matrix for chondrocyte culture are PGA, PLA, and PLAGA [123]. Among these, PGA showed better cellular attachment, characteristic chondrocytic morphology, and produced higher sulfated GAGs during in vitro culture. Hybrid scaffolds of these synthetic biodegradable polymers with natural polymers such as poly-L-lysine, agarose, and fibrin glue showed significant improvements in chondrocyte

attachment [124]. Multiphase implants mimicking different phases of cartilage such as the articulating surface phase, porous cartilage phase, and subchondral bone phase have been developed from biodegradable PGA/PLAGA,
which showed a high percentage of hyaline cartilage formation and neo-
cartilage integration in vivo [125]. Photopolymerizable injectable hydrogels
based on polyethylene oxide have been developed as an alternative to polymer
scaffolds for cartilage tissue engineering [126]. The natural biodegradable
polymers investigated for cartilage tissue engineering include chitosan, collagen, alginic acid, agarose, and hyaluronic acid [85, 127–130].

3.2.5
Bone

Bone is a complex, highly organized living organ forming the structural
framework of the body and is composed of an inorganic mineral phase of hydroxyapatite (60% by weight) and an organic phase of mainly type I collagen.
Conventional therapies for bone injuries caused by disease or trauma include
surgical reconstruction, transplantation, and artificial prosthesis. Tissue engineering has developed into an alternative therapy to treat bone loss. As in
the case of other organs, tissue engineering of bone requires cellular components, mainly osteoblasts, a 3-D scaffold for the attachment, proliferation, and
differentiation of osteoblast cells, and growth factors that modulate cellular
growth and differentiation. The primary criterion for selecting materials for
bone tissue engineering is that they should be osteocompatible. In addition,
the polymeric materials should have high moldability so that they can be processed into porous scaffolds which could allow nutrient and waste diffusion.
The polymeric materials should also exhibit mechanical properties similar to
those of native bone.

Several synthetic and natural biodegradable polymers have been investigated as matrices for bone tissue engineering. Further, the composites of
these polymers with inorganic minerals such as hydroxyapatite have been
extensively investigated [131]. Among synthetic biodegradable polymers,
aliphatic polyesters (PLA/PLAGA) have been extensively investigated for
bone tissue engineering applications. These polymers are known to show
good osteocompatibility and adhered cells are shown to retain their phenotype [132–134]. Several fabrication methods have been developed to form
3-D matrices from these polymers such as particulate leaching [135], sintered
microsphere technology [136], and soft lithography [137]. Further, PLAGA
matrices have been investigated as matrices for releasing growth factors such
as bone morphogenic proteins to accelerate bone regeneration [138]. Another
aliphatic polyester currently under investigation for bone tissue engineering
applications is PCL [139]. However, the bulk degradation profile of PLAGA
raises concerns over its use as a scaffolding material for bone tissue engineering, since it can lead to catastrophic loss in mechanical properties resulting

in failure of the implant. Hence, surface-eroding polymers such as polyanhydrides have been investigated as matrices for bone tissue engineering [140, 141]. However, poly(anhydride-*co*-imide)s seemed to be a better candidate for bone tissue engineering due to their better mechanical properties compared to those of polyanhydrides [142, 143]. The high osteocompatibility of poly(anhydride-*co*-imide)s have been established by in vitro and in vivo evaluations [143–146]. Other polymers investigated for bone tissue engineering include polyphosphazenes, polyphosphoesters, polycarbonates, poly(imino carbonates) and poly(ester urethanes). Natural polymers such as collagen, hyaluronic acid, and chitosan are also being investigated as suitable matrices mainly for low load-bearing applications.

4
Biodegradable Polymers in Drug Delivery

Controlled drug delivery technology has now emerged as a truly interdisciplinary science aimed at improving human health. The basic goal of a controlled drug delivery system is to deliver a biologically active molecule at a desired rate for a desired duration, so as to maintain the drug level in the body within the therapeutic window. Most of the initially developed drug delivery systems were based on nondegradable polymers. However, the advent of synthetic biodegradable polymers, coupled with the fact that macromolecules such as proteins can be effectively delivered from these matrices, gave a new impetus to this branch of science.

4.1
Scope of Controlled Drug Delivery

Conventionally, drugs are administered into the body via an oral or intravenous route. In these modes of administration, the therapeutic concentration of a drug is maintained in the body by repeated drug administration. The problem with such an approach is that during each repeated administration, the concentration of drug in the body peaks and declines rapidly, particularly when the elimination rate of drug is very high. This can lead to an unfavorable situation where at certain times the drug concentration in the body will be too low to provide therapeutic benefit, and at other times the concentration will be too high resulting in adverse side effects. This is particularly of great concern when using very active biotechnological drugs with narrow therapeutic windows. Moreover, this approach shows low compliance by the patient.

Controlled drug delivery technology has developed into an attractive modality to solve many of the problems of traditional drug administration by regulating the rate as well as the spatial localization of therapeutic agents [147]. In controlled drug delivery, usually the active agent will be entrapped in an insoluble matrix from which the agents will be released in a controlled fashion. Furthermore, the release characteristics of the active agents can be effectively controlled by suitably engineering the matrix parameters.

Controlled drug delivery systems can be broadly classified into temporal and targeted drug delivery systems [148, 149]. Temporal drug delivery systems are designed to release therapeutic levels of drugs from a matrix for the desired period of time. The advantage of such a system is that the therapeutic concentration of a drug can be maintained in the body for longer times without repeated administration, thereby eliminating the problems of drug under- or overdosage. Furthermore, it is more economical due to lower drug wastage, reproducible, and it increases patient compliance [150]. Attempts are still being made to develop novel delivery systems that can further fine-control the drug release pattern by synthesizing novel polymers as matrix systems, as well as by developing smart systems that can deliver multiple drugs in a controlled manner or under the effect of an external stimulus [151–153]. Targeted drug delivery systems are designed to deliver drugs at the proper dosage for the required amount of time to a specific site of the body where it is needed, thereby preventing any adverse effects drugs may have on other organs or tissues [154]. Targeted delivery assumes great importance particularly in the case of highly toxic drugs such as chemotherapeutic drugs and highly active and fragile biotechnological molecules such as peptides and proteins. Furthermore, targeted drug delivery systems can be employed to deliver drugs to sites that are inaccessible under normal conditions such as the brain. Depending on the mechanisms of action, targeted drug delivery systems can be classified into passive targeting and active targeting. Passive targeting makes use of delivery vehicles such as liposomes or micelles. These structures, due to their optimal size, can target cancer tissues or inflammatory sites with leaky vasculature as well as pass the blood–brain barrier. Active targeting makes use of the above-mentioned delivery vehicles or polymer–drug conjugates with a targeting agent like an antibody which can target a biomarker such as an antigen on a tissue or organ.

4.2
Polymeric Drug Delivery Systems

Based on the nature of the carrier, controlled drug delivery systems can be broadly classified into liposomal, electromechanical, and polymeric delivery systems [155–157]. The liposomal and infusion-pump delivery systems hold promise as drug delivery systems. However, the focus of this section is polymeric drug delivery systems.

In polymeric drug delivery systems the drugs are incorporated in a polymer matrix. The rate of release of drugs from such a system depends on a multitude of parameters such as nature of the polymer matrix, matrix geometry, properties of the drug, initial drug loading, and drug–matrix interaction [157, 158]. Extensive work has gone into optimizing these parameters to obtain the desired drug release profiles from polymeric matrices. The mechanism of release of drugs from these matrices can be controlled by physical or chemical means [155, 157]. The physically controlled release mechanisms can be classified into diffusion- and solvent-controlled systems. The chemically controlled release mechanisms can be obtained by dispersing drugs in a biodegradable polymer matrix or by developing biodegradable polymer–drug conjugates.

Most of the polymers initially investigated for drug delivery applications were hydrophobic, nondegradable polymers such as poly(dimethylsiloxane) (PS), polyurethanes (PU), and poly(ethylene-co-vinyl acetate) (EVA) [157, 159]. However, the in vivo applications of nondegradable polymer-based delivery systems are severely limited as they require a second surgical procedure to remove the implant. Thus, the nondegradable polymers are commercially developed as delivery matrices, in applications where the removal of the implants is easy. Some of the commercially developed delivery systems based on nondegradable polymers are Ocusert (a reservoir system based on EVA), Norplant (a reservoir system based on PS), MEDIFILM-400 (a transdermal patch based on PU), and TransdermScop (a transdermal patch based on polyesters). Since biodegradable polymers do not have to be removed after implantation, they became attractive candidates for drug delivery applications. Moreover, they are well-suited as carrier matrices for macromolecular drugs where the drugs can be released effectively by bioerosion of the matrices. Thus, both natural and synthetic biodegradable polymers have been extensively investigated for drug delivery applications.

4.2.1
Applications of Biodegradable Polymers

The versatility of synthetic biodegradable polymers, where the rate of degradation of the polymers can be effectively manipulated by varying the nature and ratios of monomeric units, gave a new dimension to tune the rate of drug release from these matrices. Furthermore, the ability of many synthetic and natural biodegradable polymers to form hydrogels makes them attractive candidates for developing novel drug delivery systems.

The most extensively investigated biodegradable polymers for drug delivery applications are aliphatic polyesters such as PLA, PGA, and their copolymers (PLAGA) [160]. Initial studies on these polymers were based on monolithic formulations where two-dimensional matrices of suitable geometry were prepared by casting an organic solution of polymer and drug followed by solvent evaporation or by compression/injection molding. One of the initial studies was performed in 1970 where cyclazocine was released from PLA implants [161]. However, soon PLAGA became the polymer of choice for drug delivery due to the ease of property manipulation possible with this polymer [157, 162]. The monolithic matrices used for drug delivery applications have some intrinsic problems such as the need to be implanted in the body and also lower release rates, particularly in the case of drugs with low permeability such as steroids [155]. The release rate of these drugs from PLAGA matrices has been found to increase by using novel dosage forms having very high surface area such as microspheres or microcapsules (10–2000 μm). Further, this dosage form provides ease of administration of drug carriers via parenteral routes. Microspheres are monolithic forms where therapeutic agents are distributed throughout the polymer matrix, and microcapsules are dosage forms where therapeutic agents are encapsulated within the polymer shell [157, 163]. Polymeric microspheres can be manufactured using various techniques such as spray drying, solvent evaporation, and emulsion/solvent extraction [164–166] and microcapsules can be produced using the double emulsion technique [167, 168]. Microsphere/microcapsule-based delivery systems have been extensively used for the delivery of a number of drugs such as anesthetics, antibiotics, anti-inflammatory agents, anticancer agents, hormones, steroids, and vaccines [157, 169]. Some of the commercially developed formulations based on PLAGA microspheres are Lupron Depot, Enantone Depot, Decapeptil, and Pariodel LA.

Liposomes raise interest as drug delivery matrices due to their similarity with cell membranes and easy intravenous administration. However, the low stability of liposomes prompted the development of polymer-based submicron dosage forms such as polymeric nanospheres and nanocapsules (< 1 μm). Nanospheres are monolithic devices whereas nanocapsules are colloidal particles where therapeutic agents are encapsulated within the polymer capsule. Polymeric nanospheres are prepared mainly by the emulsion evaporation method [106] and nanocapsules by interfacial deposition–polymerization [170, 171]. Nanospheres and nanocapsules of PLAGA have been investigated as delivery forms for various drugs [172]. Another aliphatic polyester that assumes importance as a drug delivery matrix is poly(caprolactone). Due to its low degradation rate and higher permeability of drugs compared to PLAGA, PCL has been extensively investigated as a matrix for macromolecular release as well as for long-term drug delivery (Capronor) [173]. The polyesters undergo bulk degradation where the ingress of water is greater than polymer erosion. Therefore, the release of drug from appropiate matrices takes place by the combined effect of diffusion and polymer erosion. This leads to a multiphase release characteristic rather than the favorable zero-order release kinetics where the drug release rate remains constant over time. Hence, novel surface-eroding polymers such as poly(ortho esters) and polyanhydrides were developed as matrices for drug delivery which could give rise to favorable zero-order release kinetics [157]. Both monolithic implants and microspheres of poly(ortho esters) and polyanhydrides have been extensively investigated as matrices for drug delivery [174–178]. Other biodegradable polymer systems investigated as drug delivery matrices include polyphosphazenes [66], polyphosphoesters [69], poly(amino acids) [51], poly(alkyl cyanoacrylates) [48], polyhydroxy alkanoates [98, 179], and poly(glutamic acid) [99].

Hydrogel forms another attractive carrier matrix for controlled drug delivery applications, particularly for the delivery of high molecular weight protein- and peptide-based drugs. Hydrogels are hydrophilic, three-dimensional networks capable of imbibing large amounts of water or biological fluids [180]. The 3-D network in hydrogels can be created from homo- or copolymers using chemical or physical cross-linking. The physical cross-linking of polymers includes entanglement and crystallites, and chemical cross-linking includes tie-points/junctions formed by using difunctional cross-linking agents or high-energy radiation [181–183]. The advantages of hydrogel-based carrier systems include good biocompatibility, mild encapsulation techniques for drugs, and the ease of modulating drug release by carefully controlling the extent of cross-linking as well as varying the polymer structures forming the network. A variety of natural and synthetic water-soluble polymers have been investigated as drug delivery matrices [157, 180].

Almost all the systems discussed so far can give rise to controlled release of therapeutic agents so as to retain a constant drug concentration in the body for a long period of time. However, some recent investigations have revealed that such a release pattern is not fit for all therapeutic agents [184]. Therefore, a pulsed drug delivery system has been developed as an alternative technology which releases a certain amount of drug within a short period of time after a lag time. Pulsed release systems can be classified into two types: intelligent pulsed delivery where the drug delivery is controlled by an external or internal stimulus, and programmable delivery where the matrix automatically releases drug at a predetermined time [185].

One of the significant advantages of hydrogel-based drug delivery systems is that the swelling behavior and hence the rate of drug release from many of these matrices can be controlled by external stimuli such as pH, ionic strength, and temperature [186]. Ionic hydrogels formed from polymers having acidic or basic pendant groups exhibit variable swelling patterns depending on the pH of the environment. Thus, polymers having carboxyl or sulfonic acid groups such as alginic acid and hyaluronic acid show a greater swelling property due to ionization of the pendant groups when the pH of the environment is greater than the pK_a of the polymer [180]. Similarly, polymers having amino groups such as chitosan show higher swelling when the pH of the environment is below the pK_a of the polymer. The pH-sensitive hydrogels have been extensively investigated as drug delivery matrices [187]. Similarly, the extent of swelling of polyelectrolyte gels is significantly influenced by the ionic strength of the swelling agent [180, 188]. Temperature-sensitive hydrogels are another class of stimulus-sensitive systems that find wide application in controlled drug delivery. Most of the initial systems developed were non-degradable and based on N-isopropyl acrylamide monomers [180]. However, recently several biodegradable thermosensitive systems based on polyesters and polyphosphazenes were developed which may find wide application in controlled release technology [189, 190].

Other than pH, ionic strength, and temperature, several other external stimuli such as light [191], magnetic field [192], ultrasound [193], electrical current [194], and the presence of some chemical [195] or biological molecules [196] can stimulate pulsatile release of drug from apropriate matrices. Another interesting pulsatile release system developed was an inflammation-induced pulsatile drug delivery system based on hyaluronic acid [197]. One of the recent advances in this field is the development of antigen [198] or protein (such as thrombin [199]) sensitive polymers. Attempts to develop multistimuli intelligent delivery systems are currently under development [200].

Programmable drug delivery systems have been developed using multi-layered polymeric matrices having different hydrophobicity/hydrophilicity as well as degradation rate. Recently, a programmable drug delivery system has been developed using a core–shell cylindrical dosage form composed of a hydrophobic coating and a cylindrical core of alternating biodegradable polyanhydride and polyphosphazene layers. A pulsatile drug release pattern can be obtained from this system based on the pH-sensitive degradation of the polymers. Further, it has been demonstrated that the lag time and duration of drug release from such a system can be modulated by varying the nature and amount of biodegradable polymers in the system [190].

5
Conclusions

As is evident from the above discussions, the availability of a wide range of biodegradable polymers of both synthetic and natural origin significantly influenced the growth of tissue engineering and controlled drug delivery technologies. However, tissue engineering is a science which is still in its infancy, with a vast potential for growth nurtured by the dramatic growth in cell and molecular biology. This produces a continuous demand for novel materials with not only appropriate physical and mechanical properties, but also suitable chemical properties to interact with the biological milieu, since tissue engineering is strongly integrated with biology. Similarly, the recent developments in controlled drug delivery technology such as targeted and stimuli-sensitive drug delivery systems also demands functionalized polymers. The future development of biodegradable polymeric biomaterials will be based on developing polymers with not only appropriate physical, mechanical, and degradation properties but also suitable biological properties.

Acknowledgements The authors acknowledge NIH (Grant #46560) for financial assistance.

References

1. Williams DF (1999) The Williams dictionary of biomaterials. Liverpool University Press, Liverpool
2. Barbucci R (ed) (2002) Integrated biomaterial science. Kluwer/Plenum, New York
3. Castner DG, Ratner BD (2002) Surf Sci 500:28
4. Okada M (2002) Prog Polym Sci 27:87
5. Kulkarni RK, Moore EG, Hegyeli AF, Leonard F (1971) J Biomed Mater Res 5:169
6. Kricheldorf HR, Kreiser-Sauders (1990) J Macromol Chem 191:1057
7. Jedlinski Z, Walach W, Kurcok P, Adamus G (1991) Macromol Chem 192:2051
8. Kricheldorf HR (2001) Chemosphere 43:49
9. Bhaw-Luximon A, Jhurry D, Spassky N, Pensec S, Belleney J (2001) Polymer 42:9651
10. Kricheldorf HR, Langanke D (2002) Polymer 43:1973
11. Middleton JC, Tipton AJ (2000) Biomaterials 21:2335
12. Katz AR, Tumer RJ (1970) Surg Gynecol Obstet 131:701
13. Ueda H, Tabata Y (2003) Adv Drug Deliv Rev 55:501
14. Griffith LG (2000) Acta Mater 48:263
15. Lewis OG, Gabisial W (1997) Sutures. In: Kirk–Othmer encyclopedia of chemical technology, 4th edn. Wiley, New York
16. Cao X, Schoichet MS (1999) Biomaterials 20:329
17. Zhu G, Mallery SR, Schwendeman SP (2000) Nat Biotechnol 18:52
18. Sabino MA, Gonzalez S, Marquez L, Feijoo JL (2000) Polym Degrad Stab 69:209
19. Cai Q, Bei J, Wang S (2002) Polymer 43:3585
20. Jeong JH, Lim DW, Han DK, Park TG (2000) Colloids Surf B Biointerfaces 18:371
21. Radder AM, Leenders H, van Blitterswijk CA (1996) J Biomed Mater Res 30:341
22. Bezemer JM, Grijpma DW, Dijkstra PJ, van Blitterswijk CA, Feijen J (2000) J Control Release 66:307
23. Schappacher M, Fabre T, Mingotaud AF, Soum A (2001) Biomaterials 22:2849
24. Arvanitoyannis I, Nakayama A, Kawasaki N, Yamamoto N (1994) Angew Makromol Chem 222:111
25. Barrows TH (1994) In: Shalaby SE (ed) Biomedical polymers designed to degrade systems. Hanser, Munich, pp 97–116
26. Edlund U, ALbertsson AC (2003) Adv Drug Deliv Rev 55:585
27. Ho LH, Huang SJ (1992) Polymer Prepr 33:94
28. Story RF, Hickey TP (1994) Polymer 35:830
29. Saad B, Hirt TD, Welti M, Uhlscgmid GK, Neuenschwander P, Suter UW (1997) J Biomed Mater Res 36:65
30. Timmer MD, Ambrose CG, Mikos AG (2003) Biomaterials 24:571
31. Payne RG, McGonigle JS, Yaszemski MJ, Yasko AW, Mikos AG (2002) Biomaterials 23:4381
32. Heller J, Barr J, Ng SY, Abdellauoi KS, Gurny R (2002) Adv Drug Deliv Rev 54:1015
33. Heller J, Barr J, Ng SY, Shen HR, Schwach-Abdellaou K, Emmah S, Rothen-Weinhold A, Gurny R (2000) Eur J Pharm Biopharm 50:121
34. Heller J, Penhale DWH, Helwing RF (1980) J Polym Sci Polym Lett 18:82
35. Ng SY, Vandamme T, Taylor MS, Heller J (1997) Macromolecules 30:770
36. Bucher JE, Slade WC (1909) J Am Chem Soc 31:1319

37. Hill JW, Carothers HW (1932) J Am Chem Soc 54:5169
38. Conix A (1958) J Polym Sci A 29:343
39. Rosen HB, Chang J, Wnek GE, Linhardt RJ, Langer R (1983) Biomaterials 4:131
40. Kumar N, Langer RS, Domb AJ (2002) Adv Drug Deliv Rev 54:889
41. Gopferich A, Tessmar J (2002) Adv Drug Deliv Rev 54:911
42. Laurencin CT, Domb A, Morris C, Brown V, Chasin M, McConnell R, Lange N, Langer R (1990) J Biomed Mater Res 24:1463
43. Uhrich KE, Ibim SEM, Larrier DR, Langer R, Laurencin CT (1998) Biomaterials 19:2045
44. Attawia MA, Uhrich KE, Botchwey E, Langer R, Laurencin CT (1996) J Orthop Res 14:445
45. Attawia MA, Uhrich KE, Botchwey E, Fan M, Langer R, Laurencin CT (1995) J Biomed Mater Res 29:1233
46. Jiang HL, Zhu KJ (2001) Biomaterials 22:211
47. Lherm C, Moiler RH, Puizieux F, Couvreur P (1992) Int J Pharm 84:13
48. Harmia T, Kreuter J, Speiser P, Boye T, Gurny R, Kubi A (1986) Int J Pharm 33:187
49. Spotnitz WD, Burks S, Mercer D (2002) Clinical indications for surgical tissue adhesives. In: Lewandrowski K-U et al. (eds) Tissue engineering and biodegradable equivalents: scientific and clinical applications. Marcel Dekker, New York, p 651
50. Anderson JM, Spilizewski KL, Hiltner K (1985) Poly(α-amino acids as biomedical polymers. In: Williams DF (ed) Biocompatibility of tissue analogs, vol I. CRC, Boca Raton, p 55
51. Bourke SL, Kohn J (2003) Adv Drug Deliv Rev 55:447
52. Kohn J, Langer R (1984) A new approach to the development of bioerodible polymers for controlled release applications employing naturally occurring amino acids. In: Polymeric materials, science and engineering, vol 51. American Chemical Society, Washington, DC, pp 119–121
53. Ertel SI, Kohn J (1994) J Biomed Mater Res 28:919
54. James K, Levene H, Parsons JR, Kohn J (1999) Biomaterials 20:2203
55. Choueka J Charvet JL, Koval KJ, Alexander H, James KS, Hooper KA, Kohn J (1996) J Biomed Mater Res 31:35
56. Pulapura S, Li C, Kohn J (1990) Biomaterials 11:666
57. Allcock HR, Kugel RL (1965) J Am Chem Soc 87:4216
58. Allcock HR, Connolly MS, Sisko JT, Al-Shali S (1988) Macromolecules 21:323
59. Scopelianos AG (1994) Polyphosphazenes as new biomaterials. In: Shalaby SW (ed) Biomedical polymers designed-to-degrade systems. Hanser, New York p 153
60. Laurencin CT, Norman ME, Elgendy HM, El-Amin SF, Allcock HR, Puscher SR, Ambrosio AA (1993) J Biomed Mater Res 27:963
61. Laurenicin CT, Ambrosio AMA, Bauer TW, Allcock HR, Attawia MA, Borden MD, Gorum WJ, Frank D (1998) The biocompatibility of polyphophazenes: evaluation in bone. Society for Biomaterials, 24th annual meeting in conjuction with 30th international symposium, April 22–26 1998, San Diego
62. Laurencin CT, El-Amin SF, Ibim SE, Willoughby DA, Attawia M, Allcock HR, Ambrosio AA (1996) J Biomed Mater Res 30:133
63. Langone F, Lora S, Veronese FM, Calceti P, Parnigotto PP, Valenti F, Palma G (1995) Biomaterials 16:347
64. Veronese FM, Marsilio F, Lora S, Caliceti P, Passi P, Orsolini P (1999) Biomaterials 20:91
65. Andrianov AK, Payne LG (1998) Adv Drug Deliv Rev 31:185
66. Lakshmi S, Katti DS, Laurencin CT (2003) Adv Drug Deliv Rev 55:467

67. Penczek S, Dida K, Kaluzynski G, Lapienis A, Nyk R, Szymanski R (1993) Makromol Chem Makromol Symp 73:91
68. Mao HQ, Shipanova-Kadiyaia I, Zhao Z, Dang W, Leong KW (1999) Biodegradable polymers: polyphosphoesters. In: Mathiowitz E (ed) Encyclopaedia of controlled drug delivery. Wiley, New York, p 45
69. Zhao Z, Wang J, Mao H, Leong KW (2003) Adv Drug Deliv Rev 55:483
70. Baeyer H, Lajous-Petter A, Debrandt W, Hampl H, Kochinke F, Herbst R (1988) J Membrane Sci 36:215
71. Hou KC, Zaniewski R, Roy S (1991) Biotechnol Appl Biochem 13:257
72. Swarbrick J, Boyan HC (1991) Gels and jellies. In: Encyclopedia of pharmaceutical technology, vol 6. Marcel Dekker, New York, p 415
73. Marques AP, Reis RL, Hunt JA (2002) Biomaterials 23:1471
74. Illum L, Fisher AN, Jabbal-Gill I, Davis SS (2001) Int J Pharm 222:109
75. Salgado AJ, Gomes ME, Chou A, Coutinho OP, Reis RL, Hutmacher DW (2002) Mater Sci Eng C 20:27
76. Horncastle J (1995) Med Device Technol 6:30
77. Mumper RJ, Huffman AS, Puolakkainen PA, Bouchard LS, Gombotz WR (1994) J Control Release 30:241
78. Gaserod O, Smidsrod O, Skjak-Braek G (1998) Biomaterials 19:1815
79. Awad HA, Erickson GR, Guilak F (2002) Material selection for engineering cartilage. In: Lewandrowski K-U et al. (eds) Tissue engineering and biodegradable equivalents: scientific and clinical applications. Marcel Dekker, New York, p 195
80. Lloyd LL, Kennedy JF, Methacanon P, Paterson M, Knill CJ (1998) Carbohydrate Polym 37:315
81. Williams D (1997) Med Device Technol 8:8
82. Khor E, Lim LY (2003) Biomaterials 24:2339
83. Ishihara M, Nakanishi K, Ono K, Sato M, Kikuchi M, Saito Y, Yura H, Mutsui T, Hattori H, Uenoyama M, Kurita A (2002) Biomaterials 23:833
84. Mi FW, Wu YB, Shyu SS, Schoung JY, Huang YB, Tsai YH, Hao JY (2002) J Biomed Mater Res 59:438
85. Lahiji A, Sohrabi A, Hungerford DS, Frondoza CG (2000) J Biomed Mater Res 51:486
86. Friess W (1998) Eur J Pharm Biopharm 45:113
87. Othman MO, Quassem W, Shahalam AP (1996) Med Eng Phys 18:584
88. Kaufman HE, Steinemann TL, Lehman E, Thompson HW, Varnell ED, Jacob-Labarre T, Gebhardt BM (1994) J Ocul Pharmacol 10:17
89. Nimni ME, Harkness RD (1988) Molecular structures and functions of collagen. In: Nimni ME (ed) Collagen vol I: biochemistry. CRC, Boca Raton p 1
90. Burke J, Yannas I, Quinby W, Bondoc C, Jung W (1981) Ann Surg 194:413
91. Royce PM, Kato T, Ohsaki K, Miura A (1995) J Dermatol Sci 10:42
92. Sung HW, Huang DM, Chang WH, Huang LL, Tsai CC, Liang IL (1999) J Biomater Sci Polym Ed 10:751
93. Tabata Y, Ikada Y (1998) Adv Drug Deliv Rev 31:287
94. Tabata Y, Nagano A, Ikada Y (1999) Tissue Eng 5:127
95. Merodio M, Irache JM, Valamanesh F, Mirshahi M (2002) Biomaterials 23:1587
96. Reddy CSK, Ghai R, Rashmi Kalia VC (2003) Bioresour Technol 87:137
97. Madison LL, Huisman GW (1999) Microbiol Mol Biol Rev 63:21
98. Kassab AC, Xu K, Denkbas EB, Dou Y, Zhao S, Piskin E (1997) J Biomater Sci Polym Ed 8:947
99. Li C, Yu DF, Newman A, Cabral F, Stephens C, Hunter N, Milas L, Wallace S (1998) Cancer Res 58:2404

100. Otani Y, Tabata Y, Ikada Y (1998) Biomaterials 19:2167
101. Otani Y, Tabata Y, Ikada Y (1998) Biomaterials 19:2091
102. Choi HJ, Yang R, Kunioka M (1995) J Appl Polym Sci 58:807
103. Fuchs JR, Nasseri BA, Vacanti JP (2001) Ann Thorac Surg 72:577
104. Laurencin CT, Ambrosio AMA, Borden MD, Cooper JA (1999) Annu Rev Biomed Eng 1:9
105. Widmer MS, Mikos AG (1998) Fabrication of biodegradable polymer scaffolds for tissue engineering. In: Patrick CW, Mikos AG, Mcintire LV (eds) Frontiers in tissue engineering. Pergamon, Oxford, p 107
106. Katti DS, Laurencin CT (2003) Synthetic biomedical polymers for tissue engineering and drug delivery. In: Shonaike GO, Advani SG (eds) Advanced polymeric biomaterials. CRC, Boca Raton, p 479
107. Leong KF, Cheah CM, Chua CK (2003) Biomaterials 24:2363
108. Meyer U, Runte C, Dirksen D, Stamm T, Fillies T, Joos U, Wiesmann HP (2003) Int Congr Ser 1256:726
109. Holbrook KA, Wolff K (1993) The structure and development of skin. In: Fitzpatrick TB, Eisen AZ, Wolff K, Freedberg IM, Austen KF (eds) Dermatology in general medicine. McGraw-Hill, New York, p 97
110. Lanza RP, Langer R, Vacanti J (2000) Principles of tissue engineering. Academic, New York
111. Falanga V, Margolis D, Alvarez O, Auletta M, Maggiacomo F, Altman M, Jensen J, Sabolinski M, Hardin-Young J (1998) Arch Dermatol 134:293
112. Langone F, Lora S, Veronese FM, Caliceti P, Parnigotto PP, Valenti F, Palma G (1995) Biomaterials 16:347
113. Yannas EV, Orgill DP, Silver J, Norregaard TV, Zervas NT, Schoene WC (1985) Trans Soc Biomater 11:146
114. Yu X, George P, Dillon GP, Bellamkonda RV (1999) Tissue Eng 5:291
115. Beaty CE, Saltzman M (1993) J Control Release 24:15
116. Plevris JN, Schina M, Haynes PC (1998) Aliment Pharmacol Ther 12:405
117. Langer R, Vacanti JP (1993) Science 260:920
118. Takeshita K, Bowen WC, Michalopoulos GK (1998) In Vitro Cell Dev Biol Anim 34:482
119. Gutomard C, Rialland L, Fremond B, Chesne C, Guillouzo A (1996) Toxicol Appl Pharmacol 141:349
120. Kawase M, Michibayashi N, Nakashima Y, Kurikawa N, Yagi K, Mizoguchi T (1997) Biol Pharm Bull 20:708
121. Griffith LG, Wu B, Cima MJ, Powers MJ, Chaignaur B, Vacanti JP (1997) Ann N Y Acad Sci 831:382
122. Takeda T, Kim TH, Lee SK, Langer R, Vacanti JP (1995) Transplant Proc 27:635
123. Freed LE, Marquis JC, Nohria A, Emmanual J, Mikos AG Langer R (1993) J Biomed Mater Res 27:11
124. Rorrer N, Aigner J, Naumann A, Planck H, Hammer C, Burmester G, Sittinger M (1998) J Biomed Mater Res 42:347
125. Niederauer GG, Slivka MA, Leatherbury NC, Korvick DL, Harroff HH, Ehler WC, Dunnn CJ, Kieswetter K (2000) Biomaterials 21:2561
126. Sims CD, Butler PE, Casanova R, Lee BT, Randolph MA, Lee WP, Vacanti CA, Yaremchuk MJ (1996) Plast Reconstruct Surg 98:843
127. Nehrer S, Breinan HA, Ramappa A, Hsu HP, Minas T, Shortkroff S, Sledge CB, Yannas IV, Spectro M (1998) Biomaterials 19:2313

128. Fragonas E, Valente M, Pozzi-Mucelli M, Toffanin R, Rizzo R, Silvestri F, Vittur F (2000) Biomaterials 21:795
129. Rahfoth B, Weisser J, Sternkopf F, Aigner T, von der Mark K, Brauer R (1998) Osteoarthr Cartil 6:50
130. Radice M, Brun P, Cortivo R, Scapinelli R, Battaliard C, Abatangelo G (2000) J Biomed Mater Res 50:101
131. Kokubo T, Kim HM, Kawashita M (2003) Biomaterials 24:2161
132. Elgendy HM, Norman ME, Keaton AR, Laurencin CT (1993) Biomaterials 14:263
133. Devin JE, Attawia MA, Laurencin CT (1996) J Biomater Sci Polym 7:661
134. Laurencin CT, Attawia MA, Elgendy HE, Herbert KM (1996) Bone 19:93S
135. Mooney DJ, Organ G, Vacanti JP, Langer R (1994) Cell Transplant 3:203
136. Borden M, Attawia M, Laurencin CT (2002) J Biomed Mater Res 61:421
137. Vozzi G, Flaim C, Ahluwalia A, Bhatia S (2003) Biomaterials 24:2533
138. Laurencin CT, Attawia MA, Lu LQ, Borden MD, Lu HH, Gorum WJ, Lieberman JR (2001) Biomaterials 22:1271
139. Ciapetti G, Ambrosio L, Savarino L, Granchi D, Cenni E, Baldini N, Pagani S, Guizzardi S, Causa F, Giunti A (2003) Biomaterials 24:3815
140. Anseth KS, Sastri VR, Langer R (1999) Nat Biotech 17:156
141. Staubli A, Mathiowitz E, Langer R (1991) Macromolecules 24:2291
142. Staubli A, Mathiowitz E, Lucarelli M, Langer R (1991) Macromolecules 24:2283
143. Attawia MA, Uhrich KE, Botchwey E, Fan M, Langer R, Laurencin CT (1995) J Biomed Mater Res 29:1233
144. Attawia MA, Herbert KM, Uhrich KE, Langer R, Laurencin CT (1999) J Biomed Mater Res B Appl Biomater 48:322
145. Attawia MA, Uhrich KE, Botchwey E, Langer R (1996) J Orthop Res 14:445
146. Ibim SE, Uhrich KE, Attawia M, Shastri VR, El-Amin SF, Bronson R, Langer R, Laurencin CT (1998) J Biomed Mater Res 43:374
147. Leong KW, Langer R (1987) Adv Drug Deliv Rev 1:199
148. Langer R (1998) Nature 392:5
149. Nimni ME (1997) Biomaterials 18:1201
150. Robinson JR, Lee VHL (eds) Controlled drug delivery: fundamentals and applications, 2nd edn. Marcel Dekker, New York
151. Qiu Y, Park K (2001) Adv Drug Deliv Rev 53:321
152. Kikuchi A, Okano T (2002) Adv Drug Deliv Rev 54:53
153. Santini JT, Cima MJ, Langer R (1999) Nature 397:335
154. Domb AJ (ed) (1994) Polymeric site-specific pharmacotherapy. Wiley, Chichester
155. Leong KW, Langer R (1987) Adv Drug Deliv Rev 1:199
156. Dash AK, Cudworth GC (1998) J Pharmacol Toxicol Methods 40:1
157. Mathiowitz E (ed) (1999) Encyclopedia of controlled drug delivery, vols I & II. Wiley, New York
158. Bae YH, Kim SW (1998) Drug delivery. In: Patrick CW, Mikos AG, Mcintire LV (eds) Frontiers in tissue engineering. Pergamon, Oxford, p 261
159. Langer R (1983) Pharmacol Ther 21:35
160. Vert M, Schwach G, Engel R, Coudane J (1998) J Control Release 53:85
161. Yolles S, Eldridge JE, Wodland JHR (1970) Polymer News 1:9
162. Lewis DH (1990) Controlled release of bioactive agents from lactide/glycolide polymers. In: Chasin M, Langer R (eds) Biodegradable polymers as drug delivery systems. Marcel Dekker, New York, pp 1–42
163. Arshady R (ed) (1999) Microspheres, microcapsules and liposomes, vol I. Preparation and chemical applications. Citrus, London

164. O'Donnell PB, McGinity JW (1997) Adv Drug Deliv Rev 28:25
165. Grattard N, Pernin M, Marty B, Roudaut G, Champion D, Le Meste M (2002) J Control Release 84:125
166. Caliceti P, Veronese FM, Lora S (2000) Int J Pharm 211:57
167. Jalil R, Nixon JR (1990) J Microencapsul 7:297
168. Iwata M, McGinity JW (1992) J Microencapsul 9:201
169. Arshady R (ed) (1999) Microspheres, microcapsules and liposomes, vol II. Medical and biotechnology applications. Citrus, London
170. Fessi H, Puisieux F, Devissaguet JP, Ammoury N, Benita S (1989) Int J Pharm 55:R1
171. Lu Z, Jianzhong B, Wang S (1999) J Control Release 61:107
172. Panyam J, Labhasetwar V (2003) Adv Drug Deliv Rev 55:329
173. Blanco MD, Bernardo MV, Sastre RL, Olmo R, Muniz E, Teijon JM (2003) Eur J Pharm Biopharm 55:229
174. Heller J, Barr J, Ng S, Shen HR, Gurny R, Schwach-Abdelaouo K, Rothen-Weinhold A, van de Weert M (2002) J Control Release 78:133
175. Heller J, Barr J, Ng SY, Schwach-Abdelaouo K, Gurny R, Vivien-Castioni N, Loup PJ, Baehni P, Momb A (2002) Biomaterials 23:4397
176. Mathiowitz E, Langer R (1987) J Control Release 5:13
177. Attawia MA, Borden MD, Herbert KM, Katti DS, Asrari F, Uhrich KE, Laurencin CT (2001) J Control Release 71:193
178. Li LC, Deng J, Stephens D (2002) Adv Drug Deliv Rev 54:963
179. Vauthier C, Dubernet C, Fattal E, Pinto-Alphandary H, Couvreur P (2003) Adv Drug Deliv Rev 55:519
180. Peppas NA, Bures P, Leobandung W, Ichikawa H (2000) Eur J Pharm Biopharm 50:27
181. Gibbs DA, Merrill EW, Smith KA, Balazs EA (1968) Biopolymers 6:777
182. Zhao L, Mitomo H, Nagasawa N, Yoshii F, Kume A (2003) Carbohydrate Polym 51:169
183. Park YD, Tirelli N, Hubbell JA (2003) Biomaterials 24:893
184. Forse RA, Mass B (1995) US Patent 005,403,590
185. Qiu LY, Zhu KJ (2001) Int J Pharm 219:151
186. Peppas NA (1991) J Bioact Compat Polym 6:241
187. Scranton AB, Rangarajan B, Klier J (1995) Adv Polym Sci 120:1
188. Ohmine I, Tanaka T (1992) J Chem Phys 77:5725
189. Jeong B, Bae YH, Kim SW (1999) Macromolecules 32:7064
190. Song S, Lee SB, Jin J, Sohn YS (1999) Macromolecules 32:2188
191. Kodzwa MG, Staben ME, Rethwisch DG (1999) J Membrane Sci 158:85
192. Zrynyi M, Szabo D, Killian HG (1998) Polym Gels Networks 6:441
193. Lavon I, Kost J (1998) J Control Release 54:1
194. Hsu CS, Block LH (1996) Pharm Res 13:1865
195. Goldbart R, Kost J (1999) Pharm Res 16:1483
196. Makino K, Mack EJ, Okano T, Kim SW (1990) J Control Release 12:235
197. Yui N, Nihira J, Okano T, Sakurai Y (1993) J Control Release 25:133
198. Miyata T, Asami N, Uragami T (1999) Nature 399:766
199. Tanihara M, Suzuki Y, Nishimura Y, Suzuki K, Kakimaru Y, Fukunishi Y (1999) J Pharm Sci 88:510
200. Kaetsu I, Uchida K, Shindo H, Gomi S, Sutani K (1999) Radiat Phys Chem 55:193

Adv Biochem Engin/Biotechnol (2006) 102: 91–111
DOI 10.1007/b138509
© Springer-Verlag Berlin Heidelberg 2005
Published online: 25 October 2005

Interface Tissue Engineering and the Formulation of Multiple-Tissue Systems

Helen H. Lu (✉) · Jie Jiang

Department of Biomedical Engineering, Fu Foundation School of Engineering
and Applied Science, Columbia University, 1210 Amsterdam Avenue,
351 Engineering Terrace Building, MC 8904, New York, NY 10027, USA
hl2052@columbia.edu

Abstract Interface tissue engineering is an exciting field which focuses on the development of tissue engineered grafts capable of promoting integration between different types of tissue and between the implant and surrounding tissue. Focusing on interface tissue engineering, and using the insertion site between the anterior cruciate ligament and bone as an example, this chapter discusses strategies in soft tissue to bone integration as well as current tissue engineering efforts in this area. This review begins with the clinical significance of this problem, followed by a review of existing fixation methods, and tissue engineering efforts aimed at addressing this critical issue. The development of multiphased scaffolds designed for the replacement of more than one type of tissue, as well as novel in vitro co-culture systems will be introduced. Future directions in the field of interface tissue engineering will also be discussed.

Keywords Co-culture · Interface tissue engineering · Ligament–bone insertion · Scaffolds

Abbreviations

3-D	Three-dimensional
ACL	Anterior crucial ligament
BG	45S5 bioactive glass
hBMSCs	Human bone marrow stromal cells
GAG	Glycosaminoglycan
PLAGA	Poly(lactide-co-glycolide)
SEM	Scanning electron microscopy

1
Introduction

A significant challenge in orthopedic tissue engineering lies in the integration of soft tissue with bone tissue. The establishment of a continuous interface is critical to the long-term success of implant systems intended for the replacement and regeneration of cartilage, ligaments and tendons. The interface or insertion connects bone and soft tissue, and its primary function is to redistribute the complex load and strains between the two types of tissue. It is also believed to act as a conduit for nutrients and cells for otherwise poorly vascularized soft tissues such as ligament or cartilage. After surgical repair or reconstruction of soft tissue and during the initial healing period, the interface between the graft and bone is mechanically the weakest point of the graft. Unfortunately the existing soft-tissue grafting systems are unable to restore both the structural and functional characteristics of the interface between bone and soft tissue.

In the past decade, tissue engineering has emerged as an alternative approach to implant design and tissue regeneration. Significant advancements have been achieved in the development of tissue engineering technologies, and several prototypes of these grafts have successfully undergone both animal and clinical trials. Design methodologies developed from current tissue engineering efforts can be readily applied to regenerate the interface between tissue types. In this review, interface tissue engineering is defined as the application of tissue engineering principles to develop scaffold systems capable of facilitating the integration between different tissue types, as well as between the biomaterial and surrounding tissue. In this chapter, both research strategies and current tissue engineering efforts in facilitating bone and soft-tissue integration will be discussed. Focusing on the regeneration of the ligament–bone interface, this chapter will describe the clinical significance of this problem, existing fixation methods, and tissue engineering efforts aimed at addressing this challenge. The tissue engineering strategies outlined in this chapter may be applied to a variety of tissue–tissue systems, with clinical relevance in the regeneration of cartilage-to-bone, tendon-to-bone, and ligament-to-bone insertions for both orthopedic and dental applications.

2
Background

2.1
Anterior Cruciate Ligament (ACL) Injuries and Reconstruction Grafts

The anterior cruciate ligament (ACL) consists of a band of regularly oriented, dense connective tissue that spans the junction between the femur and the tibia. It participates in knee motion control, acts as a joint stabilizer, and serves as the primary restraint to anterior tibial translation. The ACL is the most frequently injured knee ligament [1], and approximately 75 000 ligament repair and reconstruction procedures are performed annually in the United States [2]. Due to its intrinsically poor repair potential, the ACL does not heal upon injury and surgical intervention is often required. If untreated, injuries to the ACL will lead to functional impairment, secondary meniscus tear and the development of joint arthrosis [3, 4]. Clinically, autogenous graft based on either bone-patellar tendon-bone or hamstring tendon graft is the preferred system for ACL reconstruction. This is primarily due to a lack of alternative solutions. Synthetic ACL grafts include carbon fibers [5], Leeds–Keio ligament (polyethylene terephthalate) [6], the Gore-Tex prosthesis (poly-tetrafluoroethylene) [7], the Stryker–Dacron ligament prosthesis, which is made of Dacron tapes wrapped in a Dacron sleeve [8], and the Gore-Tex ligament augmentation device made from polypropylene [9]. These grafts have exhibited good short-term results but encounter clinical failure in the long term, as they are unable to replicate the mechanical strength and structural properties of human ACL tissue [10–12]. Limitations associated with long-term ligament repair include plastic deformation of the replacement material, weakened mechanical strength compared to the original structure and fragmentation of the replacement material due to wear [12].

Although autografts are superior to allografts, xenografts, and synthetic alternatives, ACL reconstruction based on these grafts has resulted in the loss of functional strength from the initial implantation time, followed by a gradual increase in strength that never reaches the original magnitude [13–16]. Despite its clinical success, the long-term performance of autogenous ligament substitutes is dependent on a variety of factors including the structural and material properties of the graft, the initial graft tension [17–20], the intra-articular position of the graft [21, 22], as well as graft fixation [23, 24]. Side effects such as tendonitis, arthritis, muscle atrophy and donor-site morbidity often occur. Moreover, there is often a lack of hamstring tendon graft integration with host tissue, in particular at the bony tunnels, which contributes to the suboptimal clinical outcome of these grafts [10, 11, 14]. The fixation sites at the tibial and femoral tunnels, instead of the isolated strength of the hamstring tendon graft, have been identified as the mechanically weak points in the reconstructed ACL [23, 24]. Poor interfacial integration may lead to the enlargement of the bone tunnels, and in turn compromise the long-term stability of the graft.

There is a steady rise in reported ACL injuries due to an aging and increasingly active population, exacerbated by the higher number of failures associated with current treatment modalities that require revision surgeries to correct. A disproportional number of ACL injuries occur in the teen- to middle-aged (15–35 years old) segments of the general population [2]. The relatively high level of physical activity required and desired by the active lifestyle of these individuals places extensive demands on ACL grafts, especially in terms of their fixation strength and healing potential, both immediately after surgery and during intensive rehabilitation. Furthermore, the number of revision surgeries has increased significantly in the past few years [25], and no surgical procedure has been shown to restore knee function completely without associated side-effects, such as long recovery periods, muscle atrophy, tendonitis, and arthritis.

It is clear that graft fixation is a critical weakness that severely limits the initial mechanical properties of the ligament substitutes utilized in the clinical setting. The long-term success of the reconstructed ACL is a function of the type and integrity of the initial graft fixation to host bone tissue. Thus optimized functional treatment and fixation modalities in ACL reconstruction must be developed to meet the demands of an aging yet still active population.

2.2
Current Fixation Methods Used in ACL Reconstruction

Increased emphasis has been placed on graft fixation, as the post-surgery rehabilitation protocols require the immediate ability to exercise full range of motion and reestablish neuromuscular function and weight bearing [26]. During ACL reconstruction, the bone-patellar tendon-bone or hamstring tendon graft is fixed into the tibial and femoral tunnel using a variety of fixation techniques. Fixation devices range from staples, screw and washer, press fit Endobutton®, to interference screws, accompanied by a myriad of surgical techniques for utilizing these devices. Traditionally, the bony or soft tissue is fixed within the bone tunnel or on the periosteum at a distance from the normal ligament-insertion site. The femoral fixation differs from the fixation methods utilized in the tibial insertions. The EndoButton®, and the Mitek®, anchor are utilized for the fixation of the femoral insertions, and staples, interference screws, or interferences screws combined with washers are used to fix the graft to the tibial region. The integration of quadruple semitendinosus-gracilis tendon grafts with bone is critical to the success of the indirect and direct fixation methods practised in the clinical setting.

In the past few years, the interference screw has emerged as the standard method for graft fixation. The interference screw, about 9 mm in diameter and at least 20 mm in length, is used routinely to secure tendon to bone and bone to bone in ligament reconstruction. Both metallic and polymeric interference screws have been utilized in ACL reconstruction. Surgically, the knee is flexed and the screw is inserted from the para-patellar incision into the tibial socket, and the tibial screw is screwed just underneath the joint surface. After tension is applied to the femoral graft and the knee is fully bent, the femoral tunnel screw is inserted via the anteromedial arthroscopy portal. This procedure has been reported to result in stiffness and fixation strength levels adequate for daily activities and progressive rehabilitation programs [27]. A large factor in the reported success levels associated with interference fixation can be attributed to the fact that implant fixation is now possible near the normal insertion zone [26].

While the use of interference screws has improved the fixation of ACL grafts, mechanical considerations and biomaterial-related problems associated with existing screw systems have limited the long-term functionality of the ligament substitutes [28]. Screw-related laceration of either the ligament substitute or bone plug suture has been reported [29]. In certain cases (3%), tibial screw removal was necessary to reduce the pain suffered by the patient [30]. Stress relaxation, distortion of magnetic resonance imaging, and corrosion of the metallic screws have lead to the development of biodegradable screws based on poly-α-hydroxy acids [31, 32]. A second surgery may be required to remove the metallic screws [33]. While lower incidence of graft laceration was reported for biodegradable screws [29], the highest interference fixation strength of the grafts to the tibia and femur tunnels is reported to be 475 N [34], which is significantly lower than the attachment strength of ACL to bone. When tendon-to-bone fixation with polylactic-acid-based interference screws was examined in a sheep model, intraligamentous failure was reported by six weeks [35]. Fixation strength was also found to be dependent on the quality of bone (mineral density) and bone compression. Moreover, while most biodegradable screws provide similar fixation strength as that of titanium interference screws in the fixation of bone–tendon–bone grafts [36–38], the fixation strength of degradable polymers during soft-tissue-to-bone fixation have not been fully characterized. Large-scale immune responses have also been observed for polyglycolide-based interference screws [39]. Clearly, optimal fixation of ACL grafts remains a significant clinical challenge.

2.3
Tendon-to-Bone Healing After ACL Reconstruction Surgery

Understanding the biology of tendon-to-bone healing is essential for developing an optimal rehabilitation protocol for patients who undergo ACL reconstruction surgery, as well as for the design of new fixation devices for soft-tissue-to-bone incorporation. The biochemical composition and function of the interface between tendon and bone during ACL reconstruction is poorly understood. In a study by Panni et al. [40], a persistent fibrocartilage region was only seen in the fast-healing group, suggesting that this layer may contribute to the eventual formation of direct insertions from tendon to bone. Thomopoulos et al. [41] examined matrix gene expression during healing in a rat rotator cuff-injury model. In situ hybridization studies revealed that there exists a zonal-dependent change in gene expression patterns at the insertion site. The expression levels of type I and XII collagen and aggrecan remained above normal, while the expression of collagen type X and decorin decreased over time. In the natural tendon-to-bone insertion, as in the case of the supraspinatus tendon, the zonal distribution is similar to those found in ACL to bone insertions. However, the biochemical content of the four regions may be significantly different, particularly relating to the type of collagen and matrix molecules present at the interface. The biochemical difference arises from the fact that tendons differ from ligaments in both structural and mechanical properties.

For bone-patellar tendon-bone grafts, bone-to-bone integration with the aid of interference screws is the primary mechanism facilitating graft fixation. Several groups have examined the process of tendon-to-bone healing for hamstring tendon-based ACL grafts [35, 40, 42–48]. Blickenstaff et al. evaluated the histological and biomechanical changes during the healing of a semitendinosus autograft for ACL reconstruction in a rabbit model [48]. Graft integration occurred by the formation of an indirect tendon-to-bone insertion at 26 weeks. However, large differences in graft strength and stiffness remained between the normal semitendinous tendon and ACL after 52 weeks of implantation. In a similar model, Grana et al. [47] reported that graft integration within the bone tunnel occurred by an intertwining of graft and connective tissue and anchoring of connective tissue to bone by collagenous fibers and bone formation in the tunnels. The collagenous fibers had the appearance of the Sharpey's fibers seen in an indirect tendon insertion. Rodeo et al. examined tendon-to-bone healing in a canine model by transplanting digital extensor tendon into a bone tunnel within the proximal tibial metaphysis. A layer of cellular fibrous tissue was found between the tendon and bone, and this fibrous layer matured and reorganized during the healing process. As the tendon integrated with bone through Sharpey-like fibers, the strength of the interface increased between the second and the twelfth week after surgery. The progressive increase in strength was correlated with the degree of bone ingrowth, mineralization, and maturation of the healing tissue [42].

The majority of the tendon-to-bone healing studies examined extra-articular models or fixation far away from the joint line. Panni et al. reported a dependence in the rate of graft healing (the formation of direct collagen-fiber-mediated bone–tendon junction) to the site of graft placement [40]. To approximate the original anatomy of the ACL, it is believed that fixation should be as close to the joint line as possible [49]. Recently, Weiler and associates examined tendon-to-bone healing when the graft was fixed anatomically using biodegradable poly(D,L-lactide) interference screws in a sheep model. A fibrous interface between the graft tissue and the bone tunnel was only partially developed, which was in contrast to studies in which nonanatomic fixation was used. It was reported that hamstring tendon to bone healing during compressive interference screw fixation led to partial reestablishment of the transitional zones or mineralized cartilage between soft tissue and bone at 24 weeks. Direct contact healing of the implant between the graft and the bone surface may be possible when compression is applied during healing.

The above studies have provided valuable insight into the process of tendon-to-bone healing, and have demonstrated that in-depth examination of the insertion zone is needed. It is important to note that, in most cases, tendon-to-bone healing with and without interference fixation does not result in the complete reestablishment of the normal transition zones of the native ACL–bone insertions. This inability to fully reproduce these structurally and functionally distinct regions at the junction between graft and bone is detrimental to the ability of the graft to transfer mechanical stress across the graft proper and will lead to sites of stress concentration at the junction between soft tissue and bone. A systematic characterization of the ACL–bone insertion zone will not only provide a much needed reference frame to compare tendon–bone healing, but will also facilitate the design of novel fixation devices aimed at promoting soft-tissue-to-bone healing.

3
Strategies for Interface Tissue Engineering

As discussed above, ACL injures do not heal effectively and surgical intervention is required. There is an increase in clinical utilization of hamstring tendon-based ACL graft due to the donor site morbidity associated with bone-tendon-bone grafts. Despite their distinct advantages over synthetic substitutes, autologous soft tissue grafts have a relatively high failure rate. The primary cause for the high failure rate of these grafts is the lack of consistent graft integration with the subchondral bone within the tibial and femoral tunnels. The site of tendon contact in the femoral or tibial tunnels represent the weakest point mechanically, in the early postoperative healing period [50], causing the success of ACL reconstructive surgeries to be heavily dependent on the extent of soft-tissue fixation to bone.

There has been increasing interest in finding tissue engineering solutions to soft-tissue graft to bone fixation. To develop a functional interface, several factors must be taken into consideration. First, the structural and mechanical properties of the insertion zone must be characterized. In the functional tissue engineering paradigm outlined by Butler et al. [51], the first two critical parameters determining the success of any tissue engineering effort are the determination of the material properties of the tissue to be replaced, followed by the measurement of in vivo stresses and strains in the native tissue. Neither the structural nor the mechanical properties of the insertion zone has been fully characterized. Compositional and structural distributions in the native tissue are likely correlated with functionality. Therefore, both qualitative and quantitative examinations of the interface will permit the identification and selection of the critical design parameters for scaffold design, as well as providing insight into the structure–function relationship at the interface. The ideal scaffold for the interface should be able to support the growth and differentiation of relevant cell types, while promoting the formation of multiple tissue types. The scaffold system should exhibit a gradient of structural and functional properties mimicking those of the native insertion zone. Finally, the tissue engineered graft has to be incorporated into the current design of ligament scaffolds or aid the integration of existing grafting systems for ACL repair. The following sections will review current knowledge of the structure and material properties of the ACL to bone insertion zone, as well as tissue engineering efforts focused on regenerating the soft-tissue-to-bone interface.

3.1
Structure and Biochemical Properties of the Insertion Site

Two insertion zones can be found in the human ACL: one at the femoral end and another located at the tibial attachment site. The ACL is attached to mineralized tissue through the insertion of collagen fibrils and there exists a gradual transition from soft tissue to bone. The tibial insertion zone differs structurally from the femoral insertion site, and the femoral attachment exhibits a more direct insertion of the collagen bundle into the cartilage and subchondral bone matrix. The long axis of the femoral attachment is tilted slightly forward from the vertical, and the posterior convexity is parallel to the posterior articular margin of the lateral femoral condyle. The attachment of ACL to the tibial plateau is wider than its femoral counterpart, and the ligament is inserted to the front of and lateral to the anterior tibial spine. The femoral attachment area in the human ACL was measured to be 113 ± 27 mm^2 and 136 ± 33 mm^2 for the tibia insertion [52].

Examination of morphological changes and distribution of types I and II collagen at the ACL–bone insertion sites during development will guide any reconstructional approach of the interface in vitro [53]. During development of ligament insertions in the rat knee, highly cellular ligaments insert to epiphyseal cartilage, which is in turn inserted to subchondral bone. Ossification of the epiphyseal cartilage occurs and hypertrophic chondrocytes are found near the insertion to bone. Subchondral bone formation becomes more compact and the thickness of fibrocartilage regions increases. Thus, epiphyseal cartilage resorption occurs simultaneously with osteogenesis. In addition, ligament metaplasia occurs as cartilage resorption and osteogenesis progresses, contributing to the increasing thickness of fibrocartilage during the development of ligament insertions [53, 54].

Unlike the insertion between tendon and bone, the interface between ACL and bone has not been examined in detail. It is known that, structurally, the transition from ACL to bone consists of four distinct zones: ligament, fibrocartilage, mineralized fibrocartilage, and bone [53, 55–59]. As seen in Fig. 1, the first zone, which is the ligament proper (L), is composed of solitary spindle-shaped fibroblasts aligned in rows, and is embedded in parallel collagen fibril bundles of 70–150 μm in diameter. Primarily type I collagen makes up the extracellular matrix, and type III collagen, which is composed of small reticular fibers, is located between the collagen I fibril bundles. As show in both Figs. 1 and 2, the second zone is composed of ovoid-shaped chondrocyte-like cells. The cells do not lie solitarily, but are aligned with 3–15 cells per row. Collagen fibril bundles are not strictly parallel and are much larger than those found in zone 1. Type II collagen is found within the pericellular matrix of the chondrocytes, with the matrix still predominantly consisting of type I collagen. This zone is primarily avascular and the primary sulfated proteoglycan is aggrecan. The next zone is mineralized fibrocartilage. For this region, chondrocytes appear more circular and hypertrophic, surrounded by larger pericellular matrix distal from the ACL [56]. Type X collagen, a specific marker for hypertrophic chondrocytes and subsequent mineralization, is detected and found only within this zone [55]. The inter-

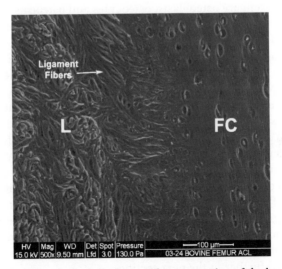

Fig. 1 Scanning electron micrograph of a sample cross section of the bovine ACL–femur insertion site, focusing on the insertion between ligament (L) to the fibrocartilage region (FC). Note the presence of collagen fibers, which directly insert into the FC region, and the ovoid chondrocytes in the fibrocartilage zone to the right. (500×)

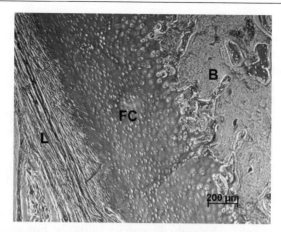

Fig. 2 Histological analyses of the bovine ligament-to-bone insertion site reveal the presence of the multiple-tissue zone and various cell types. This is an image using the modified Goldner's Masson trichrome stain of a cross section of the ACL–bone femoral insertion site. The nucleus is in black, the bone region is stained red while the soft tissue is stained green. The ligament (L), fibrocartilage (FC), and bone (B) regions can be seen. (5×)

face between mineralized fibrocartilage and subjacent bone is characterized by deep interdigitations. Increasing number of deep interdigitations is positively correlated to increased resistance to shear and tensile forces during development of rabbit ligament insertions. The last zone is subjacent bone and the cells present are osteoblasts, osteocytes and osteoclasts. The predominant collagen is type I, and fibrocartilage-specific markers such as type II collagen are no longer present.

A limited number of studies have examined the biochemical or mechanical properties and the development of each zone at the ACL–bone interface [53, 55–58, 60]. Types II, IX, and X collagen were detected within the fibrocartilaginous zone at the bovine medial collateral ligament and the ACL femoral insertion zones [55]. Type I collagen staining was reported to be lower in the insertion zone compared to the ligament proper and the bone [55]. Moreover, the distribution of type II collagen was dependent on proximity to bony areas, with type II found primarily away from the mineralized ends of the interface. Variations in collagen content in the ligament bundles are believed to be related to the differences in mechanical properties and forces experienced by these tissues [61]. While the role of fibrocartilage at this interfacial zone is not yet well understood, it may promote the integration of ligamentous tissue with bone, while responding to functional loads specific to the interface. This will be explored further in the next section.

These microscopic and qualitative examinations of the insertion zone have shed unique insight on the structural and biochemical organization of the interface. There is, however, a lack of quantitative understanding of the structural variations existing in the insertion zone, in particular in terms of the collagen distribution, collagen ratios (types I/III, I/II, I/X), fibril diameters, and cellular distribution. A systematic characterization of the interface from nanoscale to macroscale levels can yield the much needed design parameters, based on which a new generation of graft-to-bone fixation devices can be engineered. This new understanding will in turn aid the design of new ACL grafts and contribute on a broader scale to current efforts in promoting graft fixation.

3.2
Mechanical Properties of the ACL–Bone Interface

There is limited knowledge regarding the material properties of the bone–ACL insertion zones, and the specific factors determining their repair and regeneration. The above described zonal variations from soft to hard tissue at the interface are believed to facilitate a gradual change in stiffness and may prevent the build up of stress concentrations at the attachment sites [62]. However, direct measurement of the stress and strain behavior at the insertion zones has been difficult, as these regions are less than 500 μm to 1 mm in length. Consequently, there is limited data available in the literature which describes the material properties of the interface between ACL and bone. Inferred differences in material properties have been reported. Butler et al. evaluated the strain distribution within the ACL by performing failure tests of human ACL sub-bundles [63]. A spatial variation in strain was observed along the length of the ACL, with the largest strains measured at the insertion sites. In addition, it was shown that the anterior ACL sub-bundles have a significantly larger strain-energy density and failure stress values compared to the posterior ACL bundles [63]. These observations lead the authors to suggest that inhomogeneity should be introduced into the design of ligament replacement grafts.

It is well known that the weakest region between two materials of different mechanical properties is located at their interface, where the development of stress concentrations can lead to failure. When the mechanical properties of ACL were examined in a bone–ligament–bone complex, Woo et al. reported that the highest deformation occurred near or at the insertion zones [62]. The presence of a transition region comprised of fibrocartilage and mineralized fibrocartilage instead of an abrupt change from ligamentous tissue to bone in the native ACL would minimize the formation of stress concentrations in the region. In the study by Butler et al. which examined location-dependent variations in ACL mechanical properties, most of the ligaments were reported to fail at the insertion during failure tests [63]. Gao et al. reported that histological analysis revealed that avulsion fracture at or near the cement line of the subchondral bone was the most commonly observed mode of failure for ACL–bone complexes [64].

It is interesting to note that the insertion zone is dominated by non-mineralized and mineralized fibrocartilage, which are tissues adept at transmitting compressive loads. Mechanical factors may be responsible for the development and maintenance of the fibrocartilaginous zone found at many of the interfaces between soft tissue and bone [65]. The fibrocartilage zone, with its expected gradual increase in stiffness, seems to be less prone to failure [64]. It has been suggested that the fibrocartilage zone balances out the bending that otherwise would have resulted in fatigue failure [66–68]. Benjamin et al. suggested that the amount of calcified tissue in the insertion zone may be positively correlated to the force transmitted across the calcified zone [67]. Using simple histomorphometry techniques, Gao et al. determined that the thickness of the calcified fibrocartilage zone was 0.22 ± 0.7 mm and that this was not statistically different from the tibial insertion zone [54].

These observations suggest that the structure of the ACL–bone interface may be correlated with the mechanical properties and related functionality of the insertion zone. Therefore, reproducing the non-calcified and calcified fibrocartilage-rich interface in vitro on an ACL graft-fixation device may promote its integration with bone in vivo. It is expected that there will be a regional dependence of mechanical properties, varying from the ligament proper to the trabecular bone.

3.3
Design Parameters for an Interface Tissue Engineered Graft

In the past decade, tissue engineering has emerged as a possible solution to the problems associated with existing grafts for ACL reconstruction. It has the potential to provide improved clinical options through the in vitro generation of biologically based functional tissues for transplantation at the time of injury or disease. Tissue engineered ACL grafts are attractive as they exhibit the many advantages of autogenous grafts, without the associated limitations. With the advent of tissue engineering, several groups have reported on potential ACL constructs using collagen fibers, biodegradable polymers and composites. Brody et al. [69] examined the effects of canine fibroblasts seeded on knitted Dacron ligament prostheses prior to implantation. These modified prostheses demonstrated a more uniform and abundant encapsulation with connective tissue than unseeded prostheses. Dunn et al. developed skin fibroblast-seeded collagen scaffolds for ACL reconstruction [70,71], and in vivo studies found that the tissue engineered scaffolds were viable after reimplantation into the donor rabbit. Altman et al. seeded human bone marrow stromal cells (hBMSCs) on modified silk-fiber-based scaffolds with predesigned mechanical properties similar to those of human ACL [72,73]. It was reported that the hBMSCs readily differentiated in fibroblast-like cells and gene expression for type I and III collagen were up-regulated. The fiber-based scaffold geometry promoted the alignment and growth of these stem cells, and the resultant silk construct supported the growth and differentiation of bone marrow stromal cells into ligament fibroblast-like cells.

As discussed above, the interface between graft and bone is the weakest point during the initial healing period, recent efforts in ACL tissue engineering have begun to take into account the need to promote graft integration. Goulet et al. [74] developed a bioengineered ligament model, where ACL fibroblasts were added to the structure and bone plugs were used to anchor the bioengineered tissue. Fibroblasts isolated from human ACL were grown on bovine type I collagen, and the bony plugs were used to promote the anchoring of the implant within the bone tunnels. Cooper et al. [75] and Lu et al. [76] developed a tissue engineered ACL scaffold using biodegradable polymer fibers braided into a 3-D scaffold. The scaffold is comprised of three regions, one middle section with higher porosity for ligament ingrowth, and two bony attachment regions with smaller pore size and lower porosity. This scaffold has been shown to promote the attachment and growth of rabbit ACL cells in vitro and in vivo [75–77].

The identification of relevant design parameters for ligament–bone interface tissue engineering is hindered by the lack of physiologic design parameters related to the native femoral and tibial insertion zones. In-depth understanding of the structural and material properties of the native insertion zone at the nanoscale, microscale, and macroscale level is a prerequisite to formulating design parameters for a tissue engineered interface. Based on known structure–function relationships, it will be critical to mimic the architecture, as well as chemical and biological compositions of the insertion zone.

3.4
Multi-Phased Scaffold System for Interface Tissue Engineering

Scaffold design is critical in interface tissue engineering, since a supporting substrate is essential for maintaining mechanical strength, structural support, and for providing the optimal growth environment for tissue formation during the early stages of the repair process. While there are currently no reported studies directly examining the potential of multi-phased scaffolds for interface tissue engineering, our laboratory has begun to develop interfacial scaffolds aimed at regenerating the insertion site. With the native ligament–bone insertion zone as a reference point, we have formulated a multi-phased scaffold system with a gradient of chemical compositions, structural properties, and mechanical properties. Similar to the four transition zones found at the insertion site and in contrast to a homogenous scaffold, a scaffold with predesigned inhomogeneity may be able to sustain the distribution of complex stress and strain across the interfacial zones. By emulating the structural distribution of the native ligament–bone insertion zone, functional interfaces based on multi-phased scaffolds may be able to support the growth and integration of multiple-tissue systems.

Our approach is to combine a novel composite system which will support bone formation and osteointegration, with a fiber-based scaffold system which will facilitate the growth of ligamentous tissue and the formation of an interface. This composite system will be based on a 3-D composite scaffold of ceramic and biodegradable polymers. Lu et al. [78] combined poly-lactide-co-glycolide 50:50 (PLAGA) and bioactive glass (BG) to engineer a degradable, three-dimensional composite (PLAGA-BG) scaffold with improved mechanical properties. This composite is selected as the bony phase of the multi-phased scaffold proposed here (Phase C) as it has unique properties as a bone graft. The PLAGA-BG composite integrates the advantages of the parent phases, while minimizing known limitations associated with each component. A significant advantage of the composite is that it is osteointegrative. No such calcium phosphate layer is detected on PLAGA alone, and currently, osteointegration is deemed a critical factor in facilitating the chemical fixation of a biomaterial to bone tissue. Another advantage of the scaffold is that the addition of bioactive glass granules to the PLAGA matrix results in a structure with a higher compressive modulus than PLAGA alone. The compressive properties of the composite approach those of trabecular bone. Therefore, in addition to being bioactive, the PLAGA-BG would lend greater functionality in vivo compared to the PLAGA matrix alone. Moreover, the combination of the two phases serves to neutralize both the acidic byproducts produced during polymer degradation and the alkalinity due to the formation of the calcium phosphate layer. Through hydrolysis reactions, PLAGA degrades into glycolic and lactic acids, the release of which can induce a biologically significant decrease in local pH. BG releases alkaline ions which produce an elevated local pH. By forming a composite of PLAGA and BG, the acidic and basic degradation products may be neutralized, and a physiological level pH can be maintained. The composite has also been shown to support the growth and differentiation of human osteoblast-like cells in vitro [78].

This interfacial scaffold has a layered structure, with one phase optimal for ligament tissue formation and the other optimal for bone formation. The intermediate region is the region where an interfacial zone may be developed through the interaction of ACL fibroblasts and osteoblasts. We believe that a multiple-tissue system is more relevant physiologically than a scaffold with homogenous properties and may in turn promote the fixation of soft tissue to bone. By implementing the appropriate zonal-dependent variations in cell type, density, and collagen distribution into the design of multi-phased scaffolds, functional biomimetic scaffolds may be developed. By co-culturing osteoblasts and ligament fibroblasts on a multi-phased scaffold system with a gradient of material properties, we can form graft systems comprised of multiple tissues instead of just a single type of tissue. This novel scaffold system is currently been evaluated in vitro and in vivo, with promising initial results.

3.5
Development of In Vitro Co-Culture Models

Progressing through the four distinct zones which make up the native ACL insertion, several cell types can be identified, including ligament fibroblasts, chondrocytes, hypertrophic chondrocytes, osteoblasts, osteoclasts, and osteocytes. Cell-to-cell interactions may be critical in the development of a functional interface. Introduction of multiple cell types and novel co-culture systems will also be critical in facilitating the formation of the transition zones observed at the interface. To this end, the development of an in vitro multi-cell-type culture system will aid current efforts in interface tissue engineering. In addition, these model systems will augment our understanding of the developmental process of the insertion zones.

There are very few studies published in the literature describing co-culturing systems, especially in musculoskeletal systems. We first reported on an in vitro co-culture system of osteoblasts and chondrocytes, combining an osteoblast monolayer culture with a condensed micromass culture of chondrocytes [79]. It was hypothesized that osteoblast–chondrocyte interactions would lead to the development of an interfacial zone. The co-culture model permits immediate interaction between osteoblasts and chondrocytes, while maintaining the chondrogenic phenotype within the micromass. As shown in Fig. 3A, chondrocytes within the micromass exhibit spherical morphology while cells at both the surface (osteoblasts) as well as surrounding monolayer have spread. It was found that co-culture had no effect on type I collagen production by osteoblasts, but did delay mineralization. The effects of co-culture on chondrocytes were evident at the surface interaction zone, particularly in terms of glycosaminoglycan (GAG) and collagen production. Figure 3B shows that, after 14 days of culture, characteristic pericellular distribution of GAG was evident within the chondrocyte micromass region. No GAG production was observed in the osteoblastic monolayer. Co-culture and/or interactions with chondrocytes may have delayed osteoblast-mediated mineralization. The expression of specific interfacial markers such as type X collagen was confirmed in the co-cultured samples, which are preliminary confirmations of our hypothesis that co-culture may lead to the development of an interfacial zone between these cells.

Currently, there are no reported studies in the literature on neither the co-culture of ligament fibroblasts with osteoblasts, nor on the in vitro regeneration of the bone–ligament interface. Lu et al. [80] reported on initial observations of an osteoblast–ligament fibroblast co-culture. As seen in Fig. 4, after 14 days of co-culture, both human ligament fibroblasts and osteoblasts

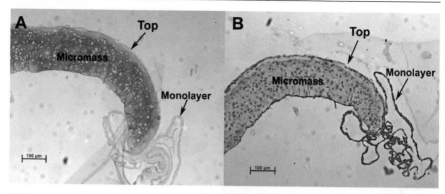

Fig. 3 Co-culture model of osteoblasts and chondrocytes. An osteoblastic monolayer is cultured on a micromass monolayer. Hematoxylin and Eosin stain (**B**) of a co-cultured micromass at day 14. Cells within the micromass exhibits spherical morphology while cells at both the surface (osteoblasts) as well as surrounding monolayer have flattened. Alcian Blue stain (**A**) of the cross section of a co-cultured micromass after 14 days of culture revealed the characteristic pericellular distribution of glycosaminoglycans (GAG) within the micromass region. No GAG production was observed in the osteoblastic monolayer. (10×)

proliferated and expanded beyond the initial seeding areas. These cells continued to grow into the interfacial zone, and eventually a contiguous and confluent culture was observed at the interface [81]. These studies demonstrate the potential of in vitro co-culture systems as a model for examining the development of an interface between ligament and fibroblasts. Results from these studies will also aid in the formulation of co-culture systems on multi-phased scaffolds.

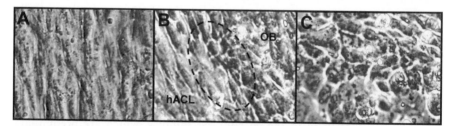

Fig. 4 Co-culture model of osteoblasts and fibroblasts. Human osteoblast-like cells and primary human ACL fibroblasts were separated by an anti-cell adhesion spacer in the culture well. The spacer was removed after 7 days and the cells were allowed to interact. It was observed that both human ACL fibroblasts (*left*) and osteoblasts (*right*) proliferated and expanded beyond the initial seeding areas after 14 days (**B**). These cells continued to grow into the interfacial zone, and eventually a contiguous and confluent culture was observed at the interface. Human ACL fibroblast (**A**) and osteoblast cultured alone (**C**) served as control groups. (32×)

3.6
In Vitro Model for Interface Tissue Engineering

Once the appropriate scaffold has been designed, it must be tested in vitro using the optimized co-culture model. By exploring the co-culture of osteoblasts, chondrocytes, and ligament fibroblasts on a multi-phased scaffold system with a gradient of material properties, graft systems comprised of multiple tissues instead of a single type of tissue may be developed. The effects of mechanical loading, growth factor and alternate cell types can be readily evaluated using this in vitro model. To this end, methodologies developed from in vitro cell co-culture models must be successfully translated onto co-cultures on biologically relevant substrates in both 2-D and 3-D forms. Recently, Spalazzi et al. [82] evaluated the interaction of bovine osteoblasts and chondrocytes on 2-D and 3-D polymer ceramic composites. As shown in Fig. 5, when a preformed osteoblast-containing matrix was present on the polymer–ceramic substrate, chondrocytes readily formed cell-matrix extensions which were absent in cultures without osteoblasts. In addition, the presence of a preformed osteoblastic layer promoted the maintenance of the spherical shape of chondrocytes, suggesting that this co-culture sys-

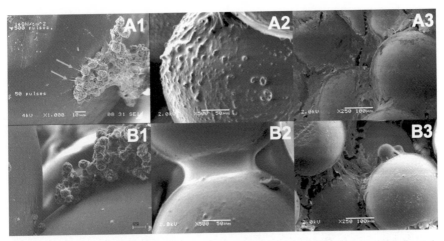

Fig. 5 Osteoblast and chondrocyte co-culture on a 3-D scaffold. Scanning electron micrographs of chondrocytes seeded on 3-D composite scaffolds in the presence (**A**) or absence (**B**) of osteoblast preformed matrix for: 1) 30 mins, 2) 24 hours, and 3) 7 days are shown here. Note that, compared to the control group (A2, 1000×), cell matrix adhesions (*arrows*) were only found on the osteoblast–chondrocyte co-culture group (A1, 1000×). Chondrocytes maintain a semi-spherical morphology on the pre-seeded scaffolds at 24 hours, as seen in A2 (500×), but are almost completely spread at the same time on the control scaffold, as seen in B2 (500×). Long-term cultures of osteoblasts–chondrocytes (A3, 250×) and chondrocytes alone (B3, 250×) revealed extensive matrix formation and coverage of the microspheres

tem on a 3-D scaffold may promote the chondrogenic phenotype in vitro. In long-term culture, it was found that more extensive cell growth and matrix elaboration were observed on the co-cultured scaffolds as compared to the chondrocyte control. It is clear that in-depth examination of cell–cell interactions during co-culture on tissue engineered scaffolds is needed, and will yield valuable information which can be utilized to optimize the interface scaffold prior to in vivo studies.

3.7
In Vivo Model for Interface Tissue Engineering

In vivo animal models for interfacial grafts will be essential for determining the healing potential of the scaffold system. Existing animal models for ligament replacement grafts are usually based on rabbit [45, 48, 66, 71, 83, 84], canine [85, 86], sheep or goat [7, 14, 87–89] systems, and they can be modified and rendered more relevant for interface tissue engineering. The development of in vivo models has not been addressed, thus significant research efforts in interface tissue engineering should be focused in this area.

4
Summary and Future Directions

Interface tissue engineering is a relatively new and exciting field with tremendous potential. The review of both background literature and current interface tissue engineering efforts presented here demonstrates that there is a pressing need for functional fixation devices capable of integrating soft-tissue grafts with bone. The clinical motivation for interface tissue engineering stems from the suboptimal performance of existing ACL reconstruction grafts and the absence of graft integration at the junction between graft and bone. There is currently a lack of in-depth understanding of the structural and mechanical properties of the ligament-to-bone insertion zones. A systematic characterization of the interface from nanoscale to macroscale levels can yield the much needed design parameters critical to the development of a new generation of graft-to-bone fixation devices. This new understanding will in turn aid in the design of new ACL grafts and contribute on a broader scale to current efforts in promoting biological graft fixation.

Moreover, the development of both co-culture systems and multi-phased scaffold systems will significantly advance existing efforts in interface tissue engineering and functional fixation devices for ACL reconstruction. This novel approach is potentially more effective since it takes into consideration the nature of the native ACL–bone insertion zone, which is comprised of distinctly ordered tissue regions including bone, mineralized and non-mineralized fibrocartilage, and ligament. By implementing the appropriate zonal-dependent variations in cell type, density, and collagen distribution into the design of multi-phased scaffolds, functional biomimetic scaffolds may be developed. The development of both in vitro and in vivo models to test the efficacy of the interfacial scaffolds will be critical for the success of the interfacial grafts. In addition, the interface tissue engineering strategies delineated here may be applied to the regeneration of cartilage-to-bone, tendon-to-bone, and ligament-to-bone insertions, with potential impact on orthopedic and dental as well as other clinical applications.

Acknowledgements The authors would like to acknowledge funding support from the Whitaker Foundation and the National Institutes of Health.

References

1. Johnson RJ (1982) Int J Sports Med 3:71–79
2. American Academy of Orthopaedic Surgeons (1997) Arthoplasty and Total Joint Replacement Procedures: United States 1990–1997, United States
3. Noyes FR, Mangine RE, Barber S (1987) Am J Sports Med 15:149–160
4. Daniel DM et al. (1994) Am J Sports Med 22:632–644
5. Thomas NP, Turner IG, Jones CB (1987) J Bone Joint Surg Br 69:312–316
6. Fujikawa K, Iseki F, Seedhom BB (1989) J Bone Joint Surg Br 71:566–570
7. Bolton CW, Bruchman WC (1985) Clin Orthop 202–213
8. McCarthy DM, Tolin BS, Schwendeman L, Friedman MJ, Woo SL (1993) The Anterior Cruciate Ligament: Current and Future Concepts, Douglas W and Jackson MD (eds.) Raven, New York
9. McPherson GK et al. (1985) Clin Orthop 186–195
10. Yahia L (1997) Ligaments and Ligamentoplasties. Springer Verlag, Berlin Heidelberg New York
11. Friedman MJ et al. (1985) Clin Orthop 9–14
12. Larson RP (1994) The Crucial Ligaments: Diagnosis and Treatment of Ligamentous Injuries About the Knee, John A, Feagin JA (eds.) Churchill Livingstone, New York
13. Jackson DW (1992) Am Acad Orthop Surg Bull 40:10–11
14. Jackson DW, Grood ES, Arnoczky SP, Butler DL, Simon TM (1987) Am J Sports Med 15:528–538
15. Jackson DW, Grood ES, Goldstein JD, Rosen MA, Kurzweil PR, Simon TM (1991) Trans Orhtop Res Soc 16:208
16. Jackson DW et al. (1993) Am J Sports Med 21:176–185
17. Fleming BC, Abate JA, Peura GD, Beynnon BD (2001) J Orthop Res 19:841–844
18. Fleming BC, Beynnon B, Howe J, McLeod W, Pope M (1992) J Orthop Res 10:177–186
19. Beynnon BD et al. (1996) J Biomech Eng 118:227–239
20. Beynnon BD et al. (1997) Am J Sports Med 25:353–359
21. Loh JC et al. (2003) Arthroscopy 19:297–304
22. Markolf KL et al. (2002) J Orthop Res 20:1016–1024
23. Kurosaka M, Yoshiya S, Andrish JT (1987) Am J Sports Med 15:225–229
24. Robertson DB, Daniel DM, Biden E (1986) Am J Sports Med 14:398–403
25. Noyes FR, Barber-Westin SD (1996) Clin Orthop 116–129

26. Brand J, Weiler A, Caborn DN, Brown CH, Johnson DL (2000) Am J Sports Med 28:761–774
27. Steiner ME, Hecker AT, Brown CH, Hayes WC (1994) Am J Sports Med 22:240–246
28. Berg EE (1996) Arthroscopy 12:232–235
29. Matthews LS, Soffer SR (1989) Arthroscopy 5:225–226
30. Kurzweil PG, Frogameni AD, Jackson DW (1995) Arthroscopy 11:289–291
31. Burkart A, Imhoff AB, Roscher E (2000) Arthroscopy 16:91–95
32. Shellock FG, Mink JH, Curtin S, Friedman MJ (1992) J Magn Reson Imaging 2:225–228
33. Safran MR, Harner CD (1996) Clin Orthop 50–64
34. Allum RL (2001) Knee 8:69–72
35. Weiler A, Hoffmann RF, Bail HJ, Rehm O, Sudkamp NP (2002) Arthroscopy 18:124–135
36. Abate JA, Fadale PD, Hulstyn MJ, Walsh WR (1998) Arthroscopy 14:278–284
37. Pena P, Grontvedt T, Brown GA, Aune AK, Engebretsen L (1996) Am J Sports Med 24:329–334
38. Weiler A, Windhagen HJ, Raschke MJ, Laumeyer A, Hoffmann RF (1998) Am J Sports Med 26:119–126
39. Weiler A, Helling HJ, Kirch U, Zirbes TK, Rehm KE (1996) J Bone Joint Surg Br 78:369–376
40. Panni AS, Milano G, Lucania L, Fabbriciani C (1997) Clin Orthop 203–212
41. Thomopoulos S et al. (2002) J Orthop Res 20:454–463
42. Rodeo SA, Arnoczky SP, Torzilli PA, Hidaka C, Warren RF (1993) J Bone Joint Surg Am 75:1795–1803
43. Liu SH et al. (1997) Clin Orthop Related Res 253–260
44. Yoshiya S, Nagano M, Kurosaka M, Muratsu H, Mizuno K (2000) Clin Orthop 278–286
45. Anderson K et al. (2001) Am J Sports Med 29:689–698
46. Chen CH et al. (2003) Arthroscopy 19:290–296
47. Grana WA, Egle DM, Mahnken R, Goodhart CW (1994) Am J Sports Med 22:344–351
48. Blickenstaff KR, Grana WA, Egle D (1997) Am J Sports Med 25:554–559
49. Fu FH, Bennett CH, Ma CB, Menetrey J, Lattermann C (2000) Am J Sports Med 28:124–130
50. Rodeo SA, Suzuki K, Deng XH, Wozney J, Warren RF (1999) Am J Sports Med 27:476–488
51. Butler DL, Goldstein SA, Guilak F (2000) J Biomech Eng 122:570–575
52. Harner CD et al. (1999) Arthroscopy 15:741–749
53. Messner K (1997) Acta Anatomica 160:261–268
54. Gao J, Messner K (1996) J Anat 188:367–373
55. Niyibizi C, Sagarrigo VC, Gibson G, Kavalkovich K (1996) Biochem Biophys Res Commun 222:584–589
56. Petersen W, Tillmann B (1999) Anat Embryol (Berl) 200:325–334
57. Sagarriga VC, Kavalkovich K, Wu J, Niyibizi C (1996) Arch Biochem Biophys 328:135–142
58. Wei X, Messner K (1996) Anat Embryol (Berl) 193:53–59
59. Cooper RR, Misol S (1970) J Bone Joint Surg Am 52:1–20
60. Clark JM, Sidles JA (1990) J Orthop Res 8:180–188
61. Mommersteeg TJ et al. (1994) J Orthop Res 12:238–245
62. Woo SL, Gomez MA, Seguchi Y, Endo CM, Akeson WH (1983) J Orthop Res 1:22–29
63. Butler DL et al. (1992) J Biomech 25:511–518

64. Gao J, Rasanen T, Persliden J, Messner K (1996) J Anat 189:127–133
65. Matyas JR, Anton MG, Shrive NG, Frank CB (1995) J Biomech 28:147–157
66. Woo SL, Newton PO, MacKenna DA, Lyon RM (1992) J Biomech 25:377–386
67. Benjamin M, Evans EJ, Rao RD, Findlay JA, Pemberton DJ (1991) J Anat 177:127–134
68. Scapinelli R, Little K (1970) J Pathol 101:85–91
69. Brody GA, Eisinger M, Arnoczky SP, Warren RF (1988) Am J Sports Med 16:203–208
70. Dunn MG, Liesch JB, Tiku ML, Maxian SH, Zawadsky JP (1994) Mater Res Soc 331:13–18
71. Bellincampi LD, Closkey RF, Prasad R, Zawadsky JP, Dunn MG (1998) J Orthop Res 16:414–420
72. Altman GH et al. (2002) Biomaterials 23:4131–4141
73. Altman GH et al. (2002) J Biomech Eng 124:742–749
74. Goulet F et al. (2000) Principles of Tissue Engineering, Lanza RP, Langer R, Vacanti JP, (eds.) Academic, New York
75. Cooper JA, Lu HH, Ko FK, Freeman JW, Laurencin CT (2005) Biomaterials 26:1523–1532
76. Lu HH, Cooper JA Jr, Manuel S, Freeman JW, Attawia MA, Ko FK, Laurencin CT (2005) Biomaterials 26:4805–4816
77. Cooper JA (2002) thesis, Drexel University
78. Lu HH, El Amin SF, Scott KD, Laurencin CT (2003) J Biomed Mater Res 64A:465–474
79. Jiang J, Nicoll SB, Lu HH (2003) Effects of Osteoblast and Chondrocyte Co-Culture on Chondrogenic and Osteoblastic Phenotype In Vitro. Trans Orth Res Soc 49
80. Lu HH, Jeffries DT, Choi RY, Oh S, Ahmad C, Policarpio EL (2003) Evaluation of Optimal Parameters in the Co-culture of Human Anterior Cruciate Ligament Fibroblasts and Osteoblasts for Interface Tissue Engineering. ASME 2003 Summer Bioengineering Conference
81. Wang IE, Jeffries DT, Jiang J, Chen FH, Lu HH (2004) Effects of co-culture on ligament fibroblast and osteoblast growth and differentiation. Transactions of the 50th Annual Meeting of the Orthopaedic Research Society
82. Spalazzi JP, Dionisio KL, Jiang J, Lu HH (2003) IEEE Eng Med Biol Mag 22:27–34
83. Amis AA, Kempson SA, Campbell JR, Miller JH (1988) J Bone Joint Surg Br 70:628–634
84. Dunn MG et al. (1992) Am J Sports Med 20:507–515
85. Bolton W, Bruchman B (1983) Aktuelle Probl Chir Orthop 26:40–51
86. Arnoczky SP, Torzilli PA, Warren RF, Allen AA (1988) Am J Sports Med 16:106–112
87. Amis AA, Camburn M, Kempson SA, Radford WJ, Stead AC (1992) J Bone Joint Surg Br 74:605–613
88. Paavolainen P, Mäkisalo S, Skutnabb K, Holmström T (1993) Acta Orthop Scand 64:323–328
89. Weiler A, Hoffmann RF, Bail HJ, Rehm O, Sudkamp NP (2002) Arthroscopy 18:124–135

Adv Biochem Engin/Biotechnol (2006) 102: 113–137
DOI 10.1007/b137207
© Springer-Verlag Berlin Heidelberg 2005
Published online: 25 October 2005

Cell Instructive Polymers

Takuya Matsumoto · David J. Mooney (✉)

Division of Engineering and Applied Sciences, Harvard University, 29 Oxford Street,
Cambridge, MA 02138, USA
tmatsu@deas.harvard.edu, mooneyd@deas.harvard.edu

Abstract Polymeric materials used in tissue engineering were initially used solely as delivery vehicles for transplanting cells. However, these materials are currently designed to actively regulate the resultant tissue structure and function. This control is achieved through spatial and temporal regulation of various cues (e.g., adhesion ligands, growth factors) provided to interacting cells from the material. These polymeric materials that control cell function and tissue formation are termed cell instructive polymers, and recent trends in their design are outlined in this chapter.

Keywords Cells · External stimuli · Growth factors · Synthetic extracellular matrix

Abbreviations

BMP	bone morphogenetic protein
CAD/CAM	computer-aided design/computer-aided manufacturing
CT	computer tomography
DDS	drug delivery system
ECM	extracellular matrix
EGF	epidermal growth factor
FDM	fused deposition molding
FGF	fibroblast growth factor
G	α-L-guluronic acid
GAG	glycosaminoglycan

HB-EGF	heparin-binding epidermal growth factor
HGF	hepatocyte growth factor
IGF	insulin-like growth factor
M	β-D-mannuronic acid
MRI	magnetic resonance imaging
mRNA	messenger RNA
PDGF	platelet-derived growth factor
PDGF-BB	platelet derived growth factor-BB
PEG	polyethylene glycol
PEO	polyethylene oxide
PET	polyethylene terephthalate
PGA	polyglycolic acid
PGCL	poly(glycolide-co-caprolactone)
PHEMA	poly(2-hydroxyethylmethacrylate)
PLA	poly lactic acid
PLG	poly(lactic acid-co-glycolic acid)
PNIPAAm	poly(N-isopropylacrylamide-co-acrylamide)
PVA	polyvinyl alcohol
SCID mice	severe combined immuno-deficiency mice
SFF	solid free-form fabrication
TGF-β	transforming growth factor-β
TRAP	tartrate-resistant acid phosphatase
VCAM-1	vascular cell adhesion molecule-1
VEGF	vascular endothelial growth factor
3D	three-dimensional

1
Introduction

A new research field in which cells are seeded into materials in order to build new biological tissues was started in the 1980's [1, 2], and this approach has since come to be recognized as one of the major strategies in the new field of tissue engineering [3]. Success in this approach to tissue engineering is based on advances in both life sciences (e.g., cell biology) and engineering (e.g., biomaterials science), as both are crucial to the strategy. Initial tissue engineering efforts typically utilized materials solely to carry cells to the desired anatomic location and/or to control the gross size and shape of the engineered tissue, and these efforts were highly successful in forming several new tissues and providing insight into the process of tissue formation [4, 5]. However, controlling this process to achieve the desired tissue structure and function, and adapting this approach to the construction of complex organs comprised of multiple cell types and tissues remain major challenges.

Tissues are typically built from several types of cells with tissue specific composition and alignment in both two and three dimensions, and contain a similarly tissue-specific extracellular matrix (ECM). For example, cells lin-ing blood vessels are aligned in monolayers, while cells in hyaline cartilage can exhibit a more complex three-dimensional organization. Similarly, the ECM in cartilage consists mainly of type II collagen, hyaluronan, and proteoglycans, while the ECM in bone consists mainly of type I collagen, along with glycoproteins and phosphoproteins. Cells typically communicate with each other as a result of several cell adhesion mechanisms [6–9], but cells in some tissues, such as corpuscular cells or chondrocytes in hyaline cartilage tissue, appear to be more independent. The cells and matrix in a tissue, furthermore, are typically dynamic. Cells change their number and phenotype during the processes of development and regeneration, and apoptosis may play an important role in these processes. The ECM produced by the cells is also changing, and these alterations may provide feed-back control over the cell fate. Altogether, these observations suggest it may be necessary to provide specific cues to regulate cell function if one desires to form complex biological tissues and regulate their function.

To successfully regulate cell fate in tissue engineering, material scientists may need to adapt new approaches to biomaterial design. In addition to providing specific physical properties, it may be necessary to provide these properties in the context of a flexible material, that can alter its properties in a pre-defined manner or in a manner dependent on the local environment. This concept differs considerably from traditional biomaterials approaches, in which a stable set of properties were typically desired. In addition, a number of cell-interactive signals may be required. To accomplish this goal, a number of molecular interactions defined by cell and molecular biologists may need to be designed into materials. We refer to these materials that control cell functions and tissue formation as cell instructive materials. In this chapter, the development of cell instructive materials and the trial of cell instructive materials for tissue engineering approaches are described. The chapter is organized into the following three sections:

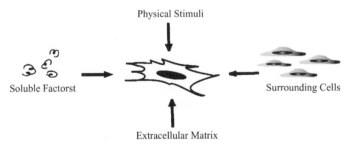

Fig. 1 A variety of environmental signals influence cell phenotype. These include soluble factors, such as growth factors, adhesive signals from the extracellular matrix, other cell populations, and physical stimuli (e.g. strain)

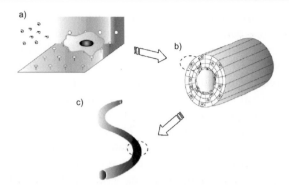

Fig. 2 Cell instructive polymers. **a** Cell instructive polymers may be used to locally provide adhesive signals or soluble factors to adherent cells. **b** These polymers can be formed into desirable three-dimensional structures. **c** The cell-polymer constructs can form new integrated tissues when placed in the body

Materials as synthetic extracellular matrices that direct cell function (Sect. 2)
The concept of designing materials to mimic the function of natural extracellular matrices, which directly regulate many cell functions, is first reviewed [10, 11]. This approach aims to regulate cell function at many levels, using materials modified either physically and/or chemically on the surface or in their bulk. The modification of materials differs in each target tissue qualitatively and quantitatively.

Designing temporally dynamic materials (Sect. 3)
Materials whose properties are designed to adjust in a pre-defined manner with time or environmental conditions may provide appropriate temporal cues to regulate cell function. Physical factors, such as temperature, pH, or mechanical stimuli may provide the environmental signals to alter the function of the materials and the cells contacting the materials [12–14]. Temporally controlled release of functional molecules from the materials can also be effective for the regulation of target tissue formation [15, 16].

Materials presenting spatially-dependent information (Sect. 4)
While most materials used to date are spatially homogeneous, providing information in a spatially defined manner may allow cell function to be differentially regulated in distinct regions of the material. Microfabrication techniques have been used to create materials that control cell position and self-assembly [17, 18], and artificially controlling specific cell arrangements may allow one to pre-define complex tissue architectures.

2
Control of Cell-Material Interactions

2.1
Mechanism of Cell and Matrix Interfaces

The strategy of preparing a composite structure, consisting of isolated cells and scaffold material, which can be used for new tissue formation both in vitro and in vivo has been widely explored in the tissue engineering field. A variety of cell functions, including proliferation, migration, and differentiation are regulated by the cellular interactions with the extracellular matrix (ECM). Therefore, to design mimics of the ECM, one must first understand the structure and function of the ECM, and the interaction between cells and the ECM. In this section, the basic mechanisms of interaction between cells and ECM molecules are described, as well as the basics of ECM constitution and cellular receptors for ECM components.

Historically, the major role of the ECM in tissues was believed to structural, in terms of providing physical stability and a space for tissue formation. However, it has become clear over the past decades that the ECM includes a number of molecules that affect activation of cell functions and organize tissue formations [19–21]. The plasma membrane of cells consists of a phospholipid bilayer, across which signals from these ECM must pass to regulate cell function. There are a variety of membrane molecules that work as receptors to the ECM in the plasma membrane, and they also can function to assemble and organize the ECM molecules secreted by the cell. Although the constituents of the ECM differ in each tissue, most tissues contain collagen, elastin, proteoglycans, and glycosaminoglycans (e.g., hyaluronan), all of which are relatively high molecular weight molecules. Moreover, a variety of glycoproteins, growth factors, matrix digesting enzymes and antagonists to these molecules are contained in the ECM. Many of these molecules regulate cellular synthesis, organization, and degradation of the ECM specific to a particular tissue, and the composition of the ECM is constantly changing during tissue formation [19–21].

Collagens constitute a large family of proteins that represent the major proteins in mammalian tissues. There are at least 21 genetically distinct types of collagens encoded by different genes [19–26]. Type I collagen is present in skin, tendon, bone, cornea and dentin, where it forms a highly crosslinked structure capable of providing tensile strength and stiffness to these tissue. Type II collagen, a major collagen in cartilage, is assembled into fibrils with minor amounts of type IX and XI collagen and crosslinked to form a co-polymer that interacts with proteoglycan [27, 28]. Basement membrane contains type IV collagen, and cornea contains type VI collagen [29, 30]. In addition to their structural functions, collagens also promote cell adhesion

Table 1 Representative heparin/heparan sulfate binding proteins

---

Growth Factors
Hepatocyte growth factor (HGF), Bone Morphogenetic Protein (BMP),
Vascular Endothelial Growth Factor (VEGF), Platelet Derived Growth Factor (PDGF),
Heparin Binding Epidermal Growth Factor (HB-EGF), Fibroblast Growth Factor (FGF)

Matrix Protein
Collagen, Fibronectin, Laminin, Vitronectin, Thrombospondin

---

as demonstrated by the DGEA (Asp-Gly-Glu-Ala) or RGD (Arg-Gly-Asp) peptides present in type I collagen, which have been identified as cell-binding domains [31, 32].

Elastin is another fibrous protein, present in many tissues, including virtually all cardiovascular and gastrointestinal track, and it provides elasticity to tissues. Although elastin is similar to collagen in terms of containing large numbers of glycine and hydroxyproline residues, it differs from collagen in that it does not include hydroxylysine. Tropoelastin, which is a precursor of elastin, is crosslinked via lysine residues to form an insoluble macromolecule with high elasticity and low turnover [21, 33, 34].

Proteoglycans are a diverse family of molecules characterized by a core protein to which is attached one or more glycosaminoglycan (GAG) sidechains. To date, more than 20 genetically different species of core proteins have been identified. The characteristic feature of GAGs is that they are composed of a disaccharide repeat sequence of two different sugars. There are several kinds of GAGs, including chondroitin sulfate, keratan sulfate, dermatan sulfate and heparan sulfate. GAGs assume an extended structure in aqueous solutions due to their strong hydrophilic nature, which is based in part on extensive sulfation. They thus hold a large number of water molecules in their molecular domain, and occupy enormous hydrodynamic space in solution as they form space filling gels. GAGs, particularly heparan sulfate and heparin have the capability to specifically interact with a number of important growth factors, and regulate cell activity via sequestration and presentation of these factors [35–38].

Hyaluronan consists of an alternating polymer of glucuronic acid and N-acetylglucosamine, joined by β1-3 linkages. Hyaluronan exists in almost all tissues and is present in large concentrations in cartilage and corpus vitreum, where it provides compressive strength due to osmotic effects. It is very large and hydrophilic, and leads to high viscosity solutions [39].

Fibronectin, laminin, vitronectin are example of glycoproteins that function in the ECM as cell adhesive molecules. These glycoproteins are typically present in the ECM or plasma at high concentrations, and adhere to cells through cell surface receptors termed integrins (described in more detail below). The RGD (Arg-Gly-Asp) amino acid sequent present in many ad-

Table 2 Representative short peptides used for surface and bulk modification of materials

Sequences	Proteins	References
RGD	Fibronectin, Type I collagen, Thrombospondin, Vitronectin	[50]
KRSR	Fibronectin	[51]
PHSRN	Fibronectin	[52]
REDV	Fibronectin	[53]
IKVAV	Laminin	[53]
YIGSR	Laminin	[54]
PDSGR	Laminin	[55]
RNIAEIIKDI	Laminin	[56]

hesive glycoproteins and ECM molecules, including fibronectin, vitronectin, osteopontin, thrombospondin, fibrinogen, von Willebrand factor and type I collagen has been extensively characterized due to its ability to serve as a cell adhesion domain [40–43]. A wide variety of other peptides present in these molecules, for example SIKVAV (Ser-Ile-Lys-Val-Ala-Val) and YIGSR (Tyr-Ile-Gly-Ser-Arg), which are present in laminin, also mediate cell adhesion [44–46]. These glycoproteins can serve as intermediates between structural ECM molecules, circulating growth factors, and cells, as they typically have several domains that bind to other molecules, such as collagen, fibrin, and heparin present in the ECM [47–49].

The first cellular ECM receptor identified was an integrin receptor [57]. Integrins are a family of membrane glycoproteins, and consist of two subunits, α and β. To date, there are 17 α subunits and 8 β subunits known. The α and β subunits, in various combinations, form at least 24 distinct integrin receptors. There are a variety of integrin combinations that are capable of binding ECM ligands. Many of these combinations, for example the fibronectin receptor ($\alpha 5\beta 1$) and vitronectin receptor ($\alpha V\beta 3$), bind to RGD sequences [58, 59]. Others, for example the laminin receptor ($\alpha 3\beta 1$, $\alpha 6\beta 1$) and collagen receptor ($\alpha 1\beta 1$), bind to non-RGD domains in molecules [60–62]. There is much overlap and redundancy in the binding of integrins to various ligands. Both subunits consist of an extracellular domain (approximately 90%), short transmembrane domain, and intracellular domain (20–60 amino acid residues) [63]. Integrins mainly bind to ECM and immunoglobulin superfamily members present extracellularly [64]. Intracellularly, integrins bind to cytoskeletal proteins and non-receptor type tyrosine kinases [65]. Thus, integrins can communicate the effects of cell-matrix interactions into the cell (outside-in signaling) and can alter the matrix in response to intracellular signals (inside-out signaling). For example, ECM signaling via integrins can induce cell morphology, migration and differentiation changes. The binding of integrins to their ligands depends on the presence of bivalent cations,

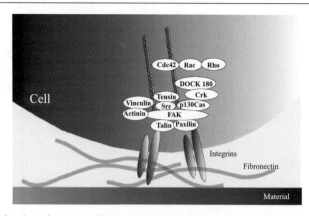

Fig. 3 Cells bind to the extracellular matrix (e.g. fibronectin) using specific transmembrane receptors termed integrins. These receptors provide a structure for the assembly of a number of extracellular structural and signaling molecules

and both active and non-active integrin receptors exist on the plasma membrane [66–69].

In addition to the integrin family, several other types of transmembrane proteoglycan serve as ECM receptors. The syndecan family contains four members (syndecan-1/syndecan, syndecan-2/fibroglycan, syndecan-3/N-syndecan, syndecan-4/ryudocan (amphyglycan)), and these receptors are transmembrane heparan sulfate proteoglycans [70, 71]. These heparan sulfate proteoglycans exhibit cell type-specific distribution, with vascular endothelial cells expressing syndecan-1, -2, and -4, and predominant targeting to basolateral surfaces. The combined transmembrane/cytoplasmic domains contain four well-conserved tyrosine residues, which are believed to serve important biological roles [72, 73]. CD44, another cell surface receptor, is a hyaluronan binding protein that consists of a single-pass transmembrane glycoprotein containing four functional domains [74, 75]. The distal extracellular domain is the region primarily responsible for the binding of hyaluronan [76]. CD44 participate in a wide variety of cellular functions, including cell-cell aggregation, retention of pericellular matrix, matrix-cell and cell-matrix signaling, receptor-mediated internalization/degradation of hyaluronan, and cell migration.

In addition to the direct regulation of cell function by ECM molecules, cell function is also dependent on the signaling of various growth factors and cytokines. Growth factors and cytokines are general terms for soluble molecules, in contrast to the typical solid-state for ECM molecules, which regulate functions such as proliferation, migration, differentiation and apoptosis. These molecules are secreted by cells, to act in an autocrine nature on the producing cells, or in a paracrine fashion on neighboring cells. Due to the central importance of these molecules in cell biology, they and their receptors and intracellular signal transduction pathways are major targets for manipulation in tissue engineering and regeneration strategies. Many wide acting factors, including fibroblast growth factor (FGF), epidermal growth factor (EGF), hepatocyte growth factor (HGF), platelet-derived growth factor (PDGF), and insulin-like growth factor (IGF) all bind to tyrosine kinase receptors. Others, including transforming growth factor-β (TGF-β), activin and bone morphogenetic protein (BMP) bind to serine-threonine kinase receptors. Growth factors and cytokines interact with the ECM in a variety of ways, which allows one to potentially intervene in cell-growth factor interactions by modulating the ECM. The ECM can serve as a reservoir for many factors, due to their binding of the factors and protection from degradation. The ECM may also be involved in growth factor presentation to cells, and in regulating expression of the factors by cells. Specific examples of these interactions include the binding of growth factors such as FGF, vascular endothelial growth factor (VEGF), heparin-binding epidermal growth factor (HB-EGF), and HGF to heparan sulfate or heparin present in the ECM to form tight complexes [77, 78]. The binding of heparan sulfate or heparin to bFGF leads to the dimerization of the growth factors, without significant conformational alternations, and these interactions can be prerequisites for binding of the growth factors to their high affinity transmembrane receptor [79, 80].

In sum, the ECM of tissues regulates cells and tissues at many levels, and cell instructive materials may be designed to mimic many aspects of the native ECM in order to regulate cellular gene expression and tissue formation.

2.2
Material Modifications to Mimic ECM Functionalities

Cells change their function through interactions with soluble mitogens, the surrounding ECM and neighboring cells, and these interactions lead to the development and regeneration of specific tissues and organs. Three-dimensional (3D) polymeric materials are often used as cell scaffolds in tissue engineering research to construct specific tissues and organs via mimicking functions of the native ECM. These artificial or synthetic ECM may allow one to alter tissue formation grossly (e.g., regulate gross structure of tissue), and regulate the microstructure of the tissue and the gene expression of the cells comprising the new tissue. In these situations, the artificial ECM is defined as a functionally bioactive material that can spatially orchestrate cell location, and temporally regulates cell phenotype. To establish an effective artificial ECM for tissue engineering, attempts have been made to directly utilize ECM molecules and/or ECM mimic molecules. This section describes several techniques by which polymer scaffolds can be modified with ECM molecules or components in an effort to control cell function and tissue formation.

Regulating the mechanism and extent of cell adhesion to synthetic ECMs is one major aspect of their design. When cells are in contact with a material that does not contain ligands for cell receptors, the relatively weak non-specific interactions between the cell and the surface (e.g., hydrophobic or electrostatic interactions) do not mediate adhesion. Cellular adhesion, and resultant control over cell function, is a result of cellular receptors binding to specific ligands presented on the surface of the material. ECM molecules may be simply adsorbed onto the material surface to provide this interaction, and adsorption of the biomacromolecules is dependent on secondary interactions, including hydrophobic and electrostatic interactions and hydrogen bonding, which often only lead to weak and readily reversible adsorption of the ECM molecules. To increase the quantity of biomacromolecules that are adsorbed on the material surface, the surface of the material may be modified physico-chemically. Increasing the surface roughness of the material by polishing or etching increases the free energy of the material surface, promoting protein adsorption and resultant cell adhesiveness to the material [81, 82]. Plasma treated polymer surfaces contain increased densities of polar functional groups, and protein adsorption and subsequent cell adhesiveness to the material can be increased significantly [83, 84].

There is considerable interest in covalently coupling cell adhesion ligands to polymeric surfaces, due to the limited ability to control protein adsorption to polymer surfaces, and the poor stability of adsorbed protein layers. Much of the work in this area builds upon the identification of small peptides that provide the cell adhesive properties of whole ECM molecules (e.g., IKVAV and YIGSR in laminin, RGD in fibronectin), as these molecules can be covalently coupled to surfaces using a variety of chemistries [85–87]. Whole ECM molecules must be isolated from a biological source, and tend to randomly fold on adsorption to the material surface so that receptor-binding domains may not always sterically available. In contrast, short peptides can be synthesized and coupled with well defined orientations. Covalent binding between functional groups by bifunctional reagents is widely used to immobilize peptides to the material [85–87]. A typical example of this approach is the covalent reaction between carboxyl groups present on polymers and the amino group on the peptide that is mediated by water-soluble carbodiimides. In many cases, several amino acids are used as a spacer to allow the cell binding component of the peptide to have sufficient mobility for efficient cell binding [88]. There are many examples of this approach being utilized to modify materials in tissue engineering. For example, aortic endothelial cells seeded on RGD-modified polylactic acid (PLA) show increased spreading compared to the same cells on unmodified PLA [86]. RGD-containing peptides covalently bound at high density to a dealdehyde starch coating on polystyrene plates enhances the adhesion, spread and growth of human umbilical vein endothelial cells compared on fibronectin-coated polystyrene [89]. Moreover, the type and quantity of peptide immobilized on the material can be readily controlled to elicit varying cell responses. Myoblast proliferation and differentiation can be regulated by varying the density of RGD ligands on the

(a)   (b)

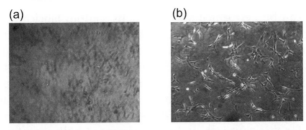

Fig. 4 RGD-modified alginate gels. Cell adhesion on the surface of alginate gel is improved by RGD immobilization. MC3T3-E1 cells show little adherence to unmodified alginate (**a**), whereas they attach and spread well on the G_4RGDY-modified alginate gel (**b**). Taken from J Dent Res [95], and used with permission of International Association of Dental Research

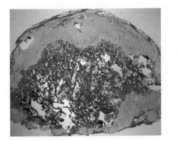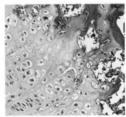

Fig. 5 Cotransplantation of osteoblasts and chondrocytes in the RGD-vehicle provided the necessary signals for the formation of growth-plate-like structures. Examination of the interface between cartilaginous and bony regions of tissue engineered with co-transplanted osteoblasts and chondrocytes demonstrated a structure similar to that seen in developing long bones. Taken from Proceedings of National Academy of Sciences, USA [96], and used with permission of National Academy of Sciences, USA

substrate surface [90]. However, the desired peptide density may be different for various desirable cell responses. For example, while initial adhesion was strongest on higher density peptide surfaces, smooth muscle cells and endothelial cells exhibited decreased matrix production on the more highly adhesive surfaces [91]. Similarly, although it has been reported that there is an increase in cell adhesion on material surface immobilized with a high-density of GRGDSY peptides, materials that immobilize peptides with lower density may lead to enhanced migration of the cells [92]. There clearly will also be a cell type dependency with these materials. For example, RGD grafted onto poly(2-hydroxyethylmethacrylate) (PHEMA) and polyethylene terephthalate (PET) led to incomplete fibroblast adhesion. In contrast, porcine aortic endothelial cells seeded on the same surface showed increased initial spreading compared to untreated polymers [93].

While many of the early attempts at material modification focused on surface modification, bulk modification may often be more relevant to tissue engineering. The processes for surface modification of materials are often simple and can be performed subsequent to scaffold fabrication. However, considering that cells are typically surrounded by the ECM in tissues, one may be able to more readily direct cell function by providing functional cues to cells interacting with the surface or in the bulk of the material, and bulk signal presentation may be especially important as the material degrades and the original surface is lost. Bulk modification of materials used to form cell encapsulating gels has been widely investigated with a number of polymers. A key issue for this approach, in both surface and bulk modification, is obtaining a high signal to noise ratio from the immobilized peptide. Materials that readily adsorb serum proteins are undesirable, as the signal provided by the peptides may rapidly become overwhelmed by uncontrolled adsorption of various proteins in a variety of conformations. We have utilized alginate gels to bypass this issue, as these naturally derived polysaccharides mediate minimal protein adsorption. RGD containing peptides have been coupled in alginate prior to gel formation, peptide presentation significantly enhances the formation of bone tissue when these gels are used to transplant bone forming populations [85, 94]. Regulating the degradation rate of the gel in concert with peptide presentation further enhances bone formation with this system [95]. Moreover, when chondrocytes and osteoblasts are cotransplanted using this material, growth plate-like tissues can be formed [96]. In a similar vein, coupling heparin to alginate allows one to stabilize growth factors and enhance their effects in vivo [97].

3
Dynamic Scaffold Materials

3.1
Preprogrammed Polymers for Factor Delivery

In the former section, the concept of combining cells and artificial extracellular matrices, and its application to tissue engineering was described. In addition to this strategy, combinations of drugs and polymer scaffold can also be an effective method for tissue engineering. The drugs commonly used in this approach are the growth factors important for chemotaxis, multiplication and differentiation of cells. The simplest approach to apply growth factors to a tissue is to simply dissolve them and introduce them into the region of the tissue defect in this solution form. However, tissue regeneration is often not found with this approach due to the rapid metabolism of the drugs in the body and their rapid degradation. To exploit growth factor signaling it may be necessary to maintain the activity and the concentration of the growth factors in the application region for an appropriate period and at the proper concentration. For this purpose, polymeric scaffold are often used to control the release of the growth factor into the tissue site at an appropriate amount and for the proper period. The polymer can protect the growth factor from degradation until its release in this approach. In the examples described in this section, the growth factor release is preprogrammed, typically due to a direct dependency of release on an environmentally insensitive degradation rate of the polymer (e.g., dependent on hydrolysis). In this situation, the release rate of the factor is intended to be equivalent in all implant sites, and all patients, irrespective of the local cellular activity at the site.

A variety of biodegradable polymers have been formed into scaffolds for the controlled release of growth factors [98, 99], and there are a number of key properties of these polymers that control their utility. The bioabsorption rate of the polymers will often regulate the rate of factor release. Degradation physically releases entrapped factor, but binding interactions between the drug and material will also regulate the availability of the drug in the surrounding fluids. In addition, similar to polymers used for cell delivery, too slow of degradation may prevent optimal tissue generation as the polymer physically prevents tissue formation.

Naturally derived polymers (e.g. collagen, gelatin, alginate, and hyaluronan) have frequently been used in tissue engineering applications because they are either components of or have macromolecular properties similar to natural ECM. Most of these naturally derived polymers are absorbed due to enzymatic action in the body. Moreover, since they are hydrophilic and absorb considerable amounts of water, oxygen and nutrients readily diffuse through these materials. Although the polymer can take various forms, such as sheets and sponges, they are commonly used in the form of hydrogels for controlled drug release applications [100, 101]. Alginate has been used in a variety of biomedical applications, including cell encapsulation and drug stabilization and delivery, because it gels under gentle conditions, has low toxicity, and is readily available. Gels are formed when divalent cations such as Ca^{2+}, Ba^{2+}, or Sr^{2+} cooperatively interact with blocks of α-L-guluronic acid (G) monomers to form ionic bridges between different polymer chains [102]. The crosslinking density and thus the mechanical properties and pore size of the ionically crosslinked gels can be manipulated by varying the β-D-mannuronic acid (M) to G ratio and the molecular weight of the polymer chain. Gels can also be formed by covalently crosslinking alginate with adipic hydrazide and polyethylene glycol (PEG) using standard carbodiimide chemistry [103, 104]. Ionically crosslinked alginate hydrogels do not specifically degrade in the body, but undergo slow, uncontrolled dissolution. Mass is lost through the ion exchange of calcium followed by the dissociation of individual chains, which results in loss of mechanical stiffness over time [105]. VEGF has been incorporated into ionically crosslinked alginate hydrogels. It is released from alginate gels both by diffusion and by mechanical stimulation [106–108]. The efficacy of this gel system to drive angiogenesis at an implant site has been demonstrated both in vitro and in vivo. Another

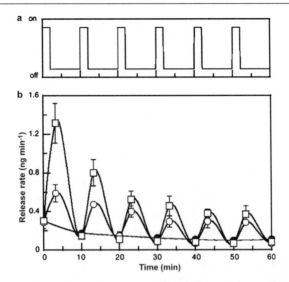

Fig. 6 A temporary dynamic strain pattern (**a**) can be used to control the release of VEGF from a polymer (**b**). Taken from Nature [107], and used with permission of Nature Publishing Group

angiogenesis-promoting protein, bFGF, has also been released in a sustained manner from heparin-alginate gels to enhance angiogenesis [109, 110].

Gelatin, denatured collagen, is also used in tissue regeneration. Importantly, the charge of gelatin is changed by the purification process used to isolate it from collagen. For example, subjecting collagen to an alkaline process results in the hydrolysis of the amide groups of the asparagine and glutamine residues, increasing the content of carboxyl groups, which makes the gelatin more negatively charged (isoelectric point is approximately 5.0). In contrast, the isoelectric point of gelatin denatured using an acidic process is approximately 9.0 [111]. For positively charged growth factors, acidic gelatin is preferable as the delivery vehicle due to the electrostatic interaction between the drug and gelatin and the resultant ability of this interaction to mediate drug release. Gelatin hydrogels have been used to immobilize bFGF, and the release of bFGF from the gel is dependent on the degradation of the gelatin hydrogel. Upon subcutaneous implantation into subcutaneous tissue mice, a bFGF-induced angiogenic effect was observed around the implantation site [112].

While naturally derived hydrogels are effective scaffolds for growth factor delivery, their mechanical strength is relatively poor. To increase the mechanical strength of hydrogels, covalent crosslinking is often utilized. Hyaluronan consists of an alternating polymer of glucuronic acid and N-acetylglucosamine joined by β1–3 linkages, and the carboxyl groups of glucuronic acid and the hydroxide group of N-acetylglucosamine serve as the main targets for chemical modification. Esterification and carbodiimide-mediated reactions are utilized for the chemical modification of the carboxyl group, and sulfation, esterification, and isourea coupling are utilized for modification of the hydroxyl group. Hyaluronan which is esterified both in the inter- and intra-molecular bonds between the hydroxyl and carboxyl groups, shows a different degradation rate and protein release profile depending on the polymer esterification degree and its molecular weight [113–115].

Synthetic polymers have also been investigated as scaffolds for bioactive molecule delivery. One of the most extensively studied types of synthetic polymers are aliphatic polyesters, such as PLA, polyglycolic acid (PGA) and poly(lactic acid-co-glycolic acid) (PLG). These polymers degrade to lactic acid and glycolic acid, by the cleavage of the ester bond resulting from hydrolysis. The bioabsorbability of these polymers can be adjusted by varying their molecular weight, crystallinity, and in the case of co-polymers, the ratio of lactide to glyolide. Several tissues have been regenerated using the controlled release of growth factors from these polymers. Bone morphogenetic protein-2 (BMP-2) delivery from porous PLG into vertebra led to a significant increase in the mechanical strength of the tissue [116]. VEGF incorporation and sustained release from porous PLG scaffolds promoted a 260% increase in the density of capillaries that invaded the matrices after 7 days of implantation [117].

A number of more sophisticated factor delivery systems have been developed that provide multiple functions or kinetics of release. Synthetic polymer microspheres are useful carriers for bioactive molecules, and can be readily combined with hydrogels or large-sized polymer scaffolds [118–120]. When microspheres are incorporated into other large-sized polymers, more complicated drug release will be available, and will also be able to attain more complex tissue generation. Blood vessels except capillaries have a complicated structure; monolayer endothelial cells are covered by mural cells, and layered smooth muscle cells with ECM. It is difficult to trigger angiogenesis and to form mature blood vessels by only single growth factor released from the scaffold. VEGF has the potential to promote angiogenesis and is concerned with endothelial cell survival. Platelet derived growth factor-BB (PDGF-BB) is a recruiting factor of mural cells or smooth muscle cells to the surrounding area of the endothelial cell monolayer and is concerned with blood vessel maturation. These two kinds of growth factor are currently incorporated into PLG matrices. Here, PDGF-BB was encapsulated in microspheres and included in a polymer scaffold, and VEGF was simply combined with a PLG polymer. It was confirmed that two kinds of growth factors were released at different manner from the scaffold polymer. When it was implanted subcutaneously in SCID mice, new blood vessels showing a relatively large size and with a mature structure were observed 4 weeks post operation [121].

As introduced here, a method of tissue regeneration has been established through the combination of various growth factors and polymer scaffold.

(a)   (b)

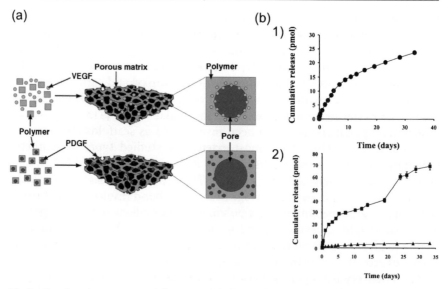

Fig. 7 Gas forming process to form growth factor delivery vehicles. Schematic of scaffold fabrication process. Growth factors were incorporated into polymer scaffolds by either mixing with polymer particles before processing into scaffolds (VEGF), or pre-encapsulating the factor (PDGF) into polymer microspheres used to form scaffolds. Typical release kinetics that can be achieved for growth factors from the scaffolds. (1) VEGF, (2) PDGF. Taken from Nature Biotechnology [121], and used with permission of Nature Publishing Group

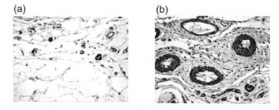

Fig. 8 Dual delivery of VEGF and PDGF leads to both an increased density of blood vessels and induces mural cell association with the vessels. Smooth muscle actin staining of tissue sections of subcutaneous implants of blank scaffolds after two weeks (**a**) and dual release of VEGF and PDGF (**b**). Taken from Nature Biotechnology [121], and used with permission of Nature Publishing Group

More sophisticated versions of this technique may allow more rapid and complex tissue regeneration. The development of advanced growth factor release technology, capable of multiple growth factor release will likely be a key component of these systems.

3.2
External Stimuli Responsive Polymers for Factor Delivery

Environmentally responsive polymer systems, in contrast to the pre-programmed systems described in the previous section, may also provide materials useful for tissue engineering. These polymers may be designed to respond to changes in temperature, mechanical cues, or other variables one desires to exploit or have regulate the regenerative process.

Temperature sensitive, injectable materials are attractive in many tissue engineering applications. Injectable materials containing cells or drugs may be placed into a tissue defect region using minimally invasive procedures, avoiding major surgery and the attendant complications to the patient. Typically, these materials have cells or drugs added while the material is dissolved in an aqueous phase, and the materials are changed to the solid phase (gel) during or after being introduced into the body. External factors, including temperature, light and pH, are used as a switch for controlling this phase change [122–126]. Thermosensitive biodegradable hydrogels consisting of blocks of polyethylene oxide (PEO) and PLA have been synthesized, and can be loaded with bioactive molecules at an elevated temperature (45 °C), and after injection the gel forms at body temperature and acts as a sustained delivery vehicle for drugs [127]. These gels have also been used to transplant chondrocytes and engineer cartilage [128]. Polypeptides can be released in a controlled manner from poly(N-isopropylacrylamide-co-acrylamide) (PNIPAAm) by exploiting the temperature-dependent gelling of these polymers [129]. Polyvinyl alcohol (PVA) has also been modified by reaction with methacrylamidoacetaldehyde dimethyl acetal to provide a crosslinkable sidechain. This modified PVA provides a photocrosslinkable hydrogel system, and when containing cell adhesion peptides has shown good cell compatibility [130].

The mechanical environment and pH of a polymer system can also be used to control drug release. Compressive loads applied to alginate hydrogels modulate the release of growth factors in a strain magnitude-dependent manner. Subcutaneous implantation of this system into mice led to a significant increase in the number of new blood vessels when periodic load was applied to the tissue [107]. A control over oligonucleotide and peptide uptake into cells can be affected using pH responsive polymers [131].

3.3
External Stimuli for Controlling Cell Phenotype

A variety of external stimuli regulate cell phenotype, and it may be necessary to incorporate certain of these into tissue engineering strategies is one desires to form fully functional tissues. Mechanical stimulation of cells comprising engineered tissue may be important in many situations, as many cell types in

the body receive a variety of mechanical stimuli. For example, the chondrocytes present in the joints of long bones are cyclically compressed by walk-
ing and the exercise inherent to daily life. Endothelial cells in blood vessels
receive the shear stress that accompanies blood flow, and tensile stress is conveyed to the smooth muscle cells. In recent years, the role and mechanism of
action of these mechanical stimuli on the phenotype of cells has been investigated both in vitro and in vivo. For examples, stretched osteoclasts enhance
messenger RNA (mRNA) expression levels of osteoclast marker enzymes,
tartrate-resistant acid phosphatase (TRAP), and cathepsin K and increase
bone-resorbing activity [132]. Static compression inhibits the proliferation of
chondrocytes and collagen types I and II gene expression in both tissue ex-
plants and collagen gel cultures. In contrast, dynamic compression stimulates
matrix synthesis and gene expression [133–136]. Shearing endothelial cells
raises the synthesis of growth factors, such as bFGF, PDGF-BB, VEGF, and
TGF-beta, and decreases vascular cell adhesion molecule-1 (VCAM-1) and oc-
cludin expression [137–139]. The mechanisms underlying these effects may
involve varying ion channel activity, regulation of cytoskeletal organization
and assembly, and control of focal adhesion formation [140, 141].

The natural regulation of cell phenotype by mechanical signals suggest
these stimuli may be used to modulate the formation and function of engineered tissues. This concept has been most thoroughly investigated in the
context of engineering vascular grafts. For example, application of an appropriate regiment of cyclic strain to smooth muscle cells in three-dimensional
scaffolds made of PLG or type I collagen induced cell alignment, expression

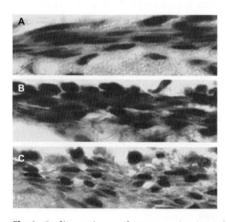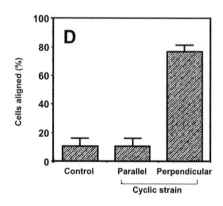

Fig. 9 Cyclic strain regulates organization of smooth muscle cells in engineered tissue. a–
c Hematoxylin-eosin stained cross sections and d quantification of cell alignment (percentage of cells aligned parallel to direction of tissue section) in smooth muscle tissues
engineered with type I collagen sponges and subjected to cyclic strain a–b or no strain
c for 10 weeks. Taken from Nature Biotechnology [142], and used with permission of
Nature Publishing Group

of matrix proteins, and an enhancement in both the ultimate tensile strength and Young's modulus of engineered tissues [142]. Similarly, flexor tendon cells within type I collagen gel aligned with the principal strain direction, and the engineered tendon had an ultimate tensile strength 3-fold greater than non-stretched engineered tendon [143]. Constant shear stress application to bone marrow stromal cells in 3D scaffolds increased calcium deposition and alkaline phosphatase activity, suggesting that osteoblastic differentiation of bone marrow stromal cells is enhanced by appropriate mechanical stimuli [144]. Oscillatory compression of chondrocyte-seeded agarose gels resulted in a dramatic increase in the equilibrium modulus, as compared with free-swelling controls [145].

Critical to exploiting mechanical stimuli in tissue engineering are the development of both bioreactors capable of applying the desired stimuli, and scaffolds that allow the mechanical signal to be conveyed to adherent cells. Systems to generate vascular networks in vitro, and stimulate with shear stress have been developed [146]. A variety of systems capable of oscillatory strain and stress application have been developed, and used to study osteogenic differentiation of cells [147]. Although materials such as collagen and PLG have been successfully used to apply mechanical loads to adherent cells, many materials may undergo plastic deformation when subjected to repeated deformation. Highly elastic biodegradable scaffolds of poly(glycolide-co-caprolactone) (PGCL) have been developed, and tensile tests these polymers can withstand extensions of 250%, and achieve over 96% recovery at these high strains [148].

4
Spatial Control of Cells Placement
and Topographical Control of Scaffold Architecture

Many tissues and organs contain functionally diverse cell types that are organized in a spatially complex arrangement. It may thus be necessary to regulate the positioning of various cell populations in engineered tissues. To achieve this aim, investigators are studying whether a tight control over scaffold architecture and cell placement can guide the morphology and positioning of cells in and on scaffolds. A thorough understanding of both normal and pathologic tissue architecture will be key to these goals.

Techniques to image and subsequently custom design tissue engineering scaffolds are under development. Crucial input in designing a scaffold architecture for a particular patient is detailed knowledge of the defect to be corrected. Imaging systems, including computer tomography (CT) and magnetic resonance imaging (MRI), are commonly used in disease diagnosis, planning medical treatments, and visualizing the region post-operation using virtual reality techniques [149–152]. More recently, it has been proposed that these imaging modalities could be used as input to custom design the gross size and morphology, and detailed internal architecture of scaffolds for tissue engineering. Computer-aided design/computer-aided manufacturing (CAD/CAM) systems can use the imaging data to create prototypes of the desired scaffolds, and techniques under development for this application include fused deposition molding (FDM) and solid free-form fabrication (SFF) [153–164].

Cell populations may be arranged and patterned in scaffolds using specific processing techniques. Regional chemical modification of substrates or controlled physical localization of cells on a chemically uniform surface may be used to micropattern cells [165–172], and these topics are the focus of other chapters in this text.

5
Conclusions

Cells and inductive molecules are the building materials for tissue engineering, and a central issue in this field is the efficient combination of these biological materials with synthetic scaffolds. Polymeric scaffolds have evolved to serve not merely as carriers of cells and inductive factors, but to actively instruct cells and provide step-by-step guidance of tissue formation. To accomplish this goal, a through understanding of the chemistry and physicochemical properties of the tissue to be engineered and the materials used in this process are required. In order to achieve the complex spatial and temporally dynamic regulation of information flow to the cells from the material may require hybrids of processing techniques and materials. Recent advances in decoding the human genome and stem cell biology will clearly impact the development of these materials in the future.

References

1. Bell E, Ehrlich HP, Buttle DJ, Nakatsuji J (1981) Science 211:1052
2. Yannas IV, Burke JF, Warpehoshi M, Stasikelis P, Skrabut EM, Orgill D, Giard DJ (1981) Am Soc Artif Intern Organs 27:19
3. Langer R, Vacanti JP (1993) Science 260:920
4. Cima LG, Vacanti JP, Vacanti C, Ingber D, Mooney D, Langer R (1991) J Biomech Eng 113:143
5. Mikos AG, Sarakinos G, Leite SM, Vacanti JP, Langer R (1993) Biomaterials 14:323
6. Madara JL (1988) Cell 53:497
7. Anderson JM, Balda MS (1993) Curr Opin Cell Biol 5:772
8. Schwarz MA, Owaribe K, Kartenbeck J, Franke WW (1990) Annu Rev Cell Biol 6:461
9. Beyrer EC (1993) Int Rev Cytol 137:1
10. Putnam AJ, Mooney DJ (1996) Nat Med 2:824
11. Kim BS, Mooney DJ (1998) Trends Biotechnol 16:224

12. Yamato M, Okuhara M, Karikusa F, Kikuchi A, Sakurai Y, Okano T (1999) J Biomed Mater Res 44:44
13. Murthy N, Campbell J, Fausto N, Hoffman A, Stayton P (2003) J Control Release 89:365
14. Kim BS, Nikolovski J, Bonadio J, Mooney DJ (1999) Nat Biotech 17:979
15. Lee KY, Peters MC, Anderson KW, Mooney DJ (2000) Nature 408:998
16. Tabata Y, Ikada Y (1998) Adv Drug Delive Rev 31:287
17. Ito Y (1999) Biomaterials 20:2333
18. Kane RS, Takayama S, Ostuni E, Ingber DE, Whitesides GM (1999) Biomaterials 20:2363
19. Hay ED (1991) Cell Biology of Extracellular Matrix, 2nd edition. Plenum Publishing New York
20. Prockop DJ, Kivirikko KI (1995) Annu Rev Biochem 64:403
21. Linda JS Charles D (1990) Extracellular matrix genes. Academic Press San Diego
22. Li K, Tamai K, Tan EM, Uitto J (1993) J Biol Chem 268:8825
23. Abe N, Muragaki Y, Yoshioka H, Inoue H, Ninomiya Y (1993) Biochem Biophys Res Commun 196:576
24. Khaleduzzaman M, Sumiyoshi H, Ueki Y, Inoguchi K, Ninomiya Y, Yoshioka H (1997) Genomics 45:304
25. Koch M, Foley JE, Hahn R, Zhou P, Burgeson RE, Gerecke DR, Gordon MK (2001) J Biol Chem 276:23120
26. Fitzgerald J, Bateman JF (2001) FEBS Lett 505:275
27. Mendler M, Eich-Bender SG, Vaughan L, Winterhalter KH, Bruckner P (1989) J Cell Biol 108:191
28. Eyre DR, Wu JJ, Fernandes RJ, Pietka TA, Weis MA (2002) Biochem Soc Trans 30:893–9
29. Johnson LD (1980) Clin Biochem 13:204
30. Zimmermann DR, Trueb B, Winterhalter KH, Witmer R, Fischer RW (1986) FEBS Lett 197:55
31. Staatz WD, Fok KF, Zutter MM, Adams SP, Rodriguez BA, Santoro SA (1991) J Biol Chem 266:7363
32. Dedhar S, Ruoslahti E, Pierschbacher MD (1987) J Cell Biol 104:585
33. Yeh H, Anderson N, Ornstein-Goldstein N, Bashir MM, Rosenbloom JC, Abrams W, Indik Z, Yoon K, Parks W, Mecham R (1989) Biochemistry 28:2365
34. Pollock J, Baule VJ, Rich CB, Ginsburg CD, Curtiss SW, Foster JA (1990) J Biol Chem 265:3697
35. Kjellen L, Lindahl U (1991) Annu Rev Biochem 60:443
36. Jackson RL, Busch SJ, Cardin AD (1991) Physiol Rev 71:481
37. Ishihara M, Guo Y, Wei Z, Yang Z, Swiedler SJ, Orellana A, Hirschberg CB (1993) J Biol Chem 268:20 091
38. Rosenberg RD, Shworak NW, Liu J, Zhang L (1997) J Clin Invest 99:2062
39. Chen WY, Abatangelo G (1999) Wound Repair Regen 7:79–89
40. Pierschbacher MD, Ruoslahti E (1984) Nature 309:30
41. Yamada KM, Kennedy DW (1984) J Cell Biol 99:29
42. Ruoslahti E, Pierschbacher MD (1987) Science 238:491
43. Lawler J, Weinstein R, Hynes RO (1988) J Cell Biol 107:2351
44. Grant DS, Kinsella JL, Fridman R, Auerbach R, Piasecki BA, Yamada Y, Zain M, Kleinman HK (1992) J Cell Physiol 153:614
45. Graf J, Ogle RC, Robey FA, Sasaki M, Martin GR, Yamada Y, Kleinman HK (1987) Biochemistry 26:6896

46. Iwamoto Y, Graf J, Sasaki M, Kleinman HK, Greatorex DR, Martin GR, Robey FA, Ya-mada Y (1988) J Cell Physiol 134:287
47. Engvall E, Ruoslahti E (1977) Int J Cancer 20:1
48. Gebb C, Hayman EG, Engvall E, Ruoslahti E (1986) J Biol Chem 261:16 698
49. Makogonenko E, Tsurupa G, Ingham K, Medved L (2002) Biochemistry 41:7907
50. Massia SP, Hubbell JA (1991) J Biomed Mater Res 25:223
51. Webster TJ, Schadler LS, Siegel RW, Bizios R (2001) Tissue Eng 7:291
52. Dillo AK, Ochsenhirt SE, McCarthy JB, Fields GB, Tirrell M (2001) Biomaterials 22:1493
53. Heilshorn SC, DiZio KA, Welsh ER, Tirrell DA (2003) Biomaterials 24:4245
54. Ranieri JP, Bellamkonda R, Bekos EJ, Vargo TG, Gardella JA, Aebischer P (1995) J Biomed Mater Res 29:779
55. Aucoin L, Griffith CM, Pleizier G, Deslandes Y, Sheardown H (2002) J Biomater Sci Polym Ed 13:447
56. Schense JC, Bloch J, Aebischer P, Hubbell JA (2000) Nat Biotechnol 18:415
57. Suzuki S, Argraves WS, Pytela R, Arai H, Krusius T, Pierschbacher MD, Ruoslahti E (1986) Proc Natl Acad Sci USA 83:8614
58. Puzon-McLaughlin W, Yednock TA, Takada Y (1996) J Biol Chem 271:16 580
59. Aznavoorian S, Stracke ML, Parsons J, McClanahan J, Liotta LA (1996) J Biol Chem 271:3247
60. Gehlsen KR, Sriramarao P, Furcht LT, Skubitz AP (1992) J Cell Biol 117:449
61. Delwel GO, de Melker AA, Hogervorst F, Jaspars LH, Fles DL, Kuikman I, Lind-blom A, Paulsson M, Timpl R, Sonnenberg A (1994) Mol Biol Cell 5:203
62. Yamamoto M, Yamato M, Aoyagi M, Yamamoto K (1995) Exp Cell Res 219:249
63. Tamkun JW, DeSimone DW, Fonda D, Patel RS, Buck C, Horwitz AF, Hynes RO (1986) Cell 46:271
64. Diamond MS, Staunton DE, Marlin SD, Springer TA (1991) Cell 65:961
65. Kornberg LJ, Earp HS, Turner CE, Prockop C, Juliano RL (1991) Proc Natl Acad Sci USA 88:8392
66. Paulsson M (1988) J Biol Chem 263:5425
67. Mould AP, Garratt AN, Puzon-McLaughlin W, Takada Y, Humphries MJ (1998) Bio-chem J 331:821
68. Takagi J, Petre BM, Walz T, Springer TA (2002) Cell 110:599
69. Takagi J, Springer TA (2002) Immunol Rev 186:141
70. Carey DJ (1997) Biochem J 327:1
71. Yoneda A, Couchman JR (2003) Matrix Biol 22:25
72. Bernfield M, Kokenyesi R, Kato M, Hinkes MT, Spring J, Gallo RL, Lose EJ (1992) Ann Rev Cell Biol 8:365
73. Rosenberg RD, Schworak NW, Lui J, Schwartz JJ, Zhang L (1997) J Clin Invest 99:2062
74. Underhill CB, Toole BP (1980) J Biol Chem 255:4544
75. Underhill CB, Green SJ, Cologlio PM, Tarone G (1987) J Biol Chem 262:13 142
76. Peach RJ, Hollenbaugh D, Stamenkovic I, Aruffo A (1993) J Cell Biol 122:257
77. Vaisman N, Gospodarowicz D, Neufeld G (1990) J Biol Chem 265:19 461
78. Folkman J, Klagsbrun M, Sasse J, Wadzinski M, Ingber D, Vlodavsky I (1988) Am J Pathol 130:393
79. Faham S, Hileman RE, Fromm JR, Linhardt RJ, Rees DC (1996) Science 271:1116
80. Ornitz DM, Herr AB, Nilsson M, Westman J, Svahn CM, Waksman G (1995) Science 268:432
81. Hallab NJ, Bundy KJ, O'Connor K, Moses RL, Jacobs JJ (2001) Tissue Eng 7:55

82. Dalby MJ, Di Silvio L, Gurav N, Annaz B, Kayser MV, Bonfield W (2002) Tissue Eng 2002 8:453
83. Olde Riekerink MB, Claase MB, Engbers GH, Grijpma DW, Feijen J (2003) J Biomed Mater Res 65A:417
84. Claase MB, Olde Riekerink MB, de Bruijn JD, Grijpma DW, Engbers GH, Feijen J (2003) Biomacromolecules 4:57
85. Rowley JA, Madlambayan G, Mooney DJ (1999) Biomaterials 20:45
86. Cook AD, Hrkach JS, Gao NN, Johnson IM, Pajvani UB, Cannizzaro SM, Langer R (1997) J Biomed Mater Res 35:513
87. Seifert P, Romaniuk P, Groth T (1997) Biomaterials 18:1495
88. Marler JJ, Guha A, Rawley J, Koka R, Mooney DJ, Upton J, Vacanti JP (2000) Plast Reconstr Surg 105:2049
89. Holland J, Hersh L, Bryhan M, Onyiriuka E, Ziegler L (1996) Biomaterials 17:2147
90. Rowley JA, Mooney DJ (2002) J Biomed Mater Res 60:217
91. Mann BK, Tsai AT, Scott-Burden T, West JL (1999) Biomaterials 20:2281
92. Neff JA, Tresco PA, Caldwell KD (1999) Biomaterials 20:2377–2393
93. Massia SP, Hubbell JA (1990) Ann NY Acad Sci 589:261
94. Alsberg E, Anderson K, Albeiruti A, Franceschi R, Mooney DJ (2001) J Dent Res 80:2025
95. Alsberg E, Kong HJ, Hirano Y, Smith MK, Albeiruti A, Mooney DJ (2003) J Dent Res 82:903
96. Alsberg E, Anderson K, Albeiruti A, Rowley J, Mooney DJ (2002) PNAS 99:12 025
97. Tanihara M, Suzuki Y, Yamamoto E, Noguchi A, Mizushima Y (2001) J Biomed Mater Res 56:216
98. Sheridan MH, Shea LD, Peters MC, Mooney DJ (2000) J Control Release 64:91
99. Yamazaki Y, Oida S, Ishihara K, Nakabayashi N (1996) J Biomed Mater Res 30:1
100. Ikada Y, Tabata Y (1998) Adv Drug Deliv Rev 31:287
101. Burdick JA, Mason MN, Hinman AD, Thorne K, Anseth KS (2002) J Control Release 83:53
102. Smidsrød O, Skjåk-Bræk G (1990) Trends Biotech 8:71
103. Eiselt P, Lee KY, Mooney DJ (1999) Macromolecules 32:5561
104. Lee KY, Rowley JA, Eiselt P, Moy EM, Bouhadir KH, Mooney DJ (2000) Macromolecules 33:4291
105. LeRoux MA, Guilak F, Setton LA (1999) J Biomed Mater Res 47:46
106. Elcin YM, Dixit V, Gitnick G (2001) Artif Organs 25:558
107. Lee KY, Peters MC, Anderson KW, Mooney DJ (2000) Nature 408:998
108. Lee KY, Peters MC, Mooney DJ (2001) Adv Mater 13:837
109. Edelman ER, Mathiowitz E, Langer R, Klagsbrun M (1991) Biomaterials 12:619
110. Edelman ER, Nugent MA, Smith LT, Karnovsky MJ (1992) J Clin Invest 89:465
111. Yamamoto M, Ikada Y, Tabata Y (2001) J Biomater Sci Polym Ed 12:77
112. Tabata Y (2003) Tiss Eng 9:s5
113. Zhong SP, Campoccia D, Doherty PJ, Williams RL, Benedetti L, Williams DF (1994) Biomaterials 15:359
114. Simon LD, Stella VJ, Charman WN, Charman SA (1999) J Control Release 61:267
115. Prestwich GD, Marecak DM, Marecek JF, Vercruysse KP, Ziebell MR (1998) J Control Release 53:93
116. Muscher GF, Hyodo A, Manning T, Kambic H, Easley K (1994) Clin Orthop 308:229
117. Peters MC, Polverini PJ, Mooney DJ (2002) J Biomed Mater Res 60:668
118. Richardson TP, Murphy WL, Mooney DJ (2003) Plast Reconstr Surg 112:162
119. Berkland C, King M, Cox A, Kim K, Pack DW (2002) J Control Release 82:137

120. Zellin G, Linde A (1997) J Biomed Mater Res 35:181
121. Richardson TP, Peters MC, Ennett AB, Mooney DJ (2001) Nature Biotech 19:1029
122. Elisseeff J, McIntosh W, Fu K, Blunk BT, Langer R (2001) J Orthop Res 19:1098
123. Jeong B, Gutowska A (2002) Trends Biotechnol 20:305
124. Nguyen KT, West LJ (2002) Biomaterials 23:4307
125. Shimoboji T, Ding ZL, Stayton PS, Hoffman AS (2002) Bioconjugate chem
126. Ibusuki S, Fujii Y, Iwamoto Y, Matsuda T (2003) Tissue Eng 9:371
127. Jeong B, Bae YH, Lee DS, Kim SW (1997) Nature 388:860
128. Jeong B, Lee KM, Gutowska A, An YH (2002) Biomacromolecules 3:865
129. Chilkoti A, Dreher MR, Meyer DE, Raucher D (2002) Advance Drug Deliv Rev 54:613
130. Schmedlen RH, Masters KS, West JL (2002) Biomaterials 23:4325
131. Murthy N, Campbell J, Fausto N, Hoffman AS, Stayton PS(2003) Bioconjugate Chem
     14:412
132. Kurata K, Uemura T, Nemoto A, Tateishi T, Murakami Y, Higaki H, Miura H, Iwa-
     moto Y (2001) J Bone Miner Res 16:722
133. Kim YJ, Grodzinsky AJ, Plaas AH (1996) Arch Biochem Biophys 328:331
134. Quinn TM, Grodzinsky AJ, Buschmann MD, Kim YJ, Hunziker EB (1998) J Cell Sci
     111:573
135. Hunter CJ, Imler SM, Malaviya P, Nerem RM, Levenston ME (2002) Biomaterials
     23:1249
136. Ragan PM, Badger AM, Cook M, Chin VI, Gowen M, Grodzinsky AJ (1999) J Orthop
     Res 17:836
137. Conklin BS, Zhong DS, Zhao W, Lin PH, Chen C (2002) J Surg Res 102:13
138. Malek AM, Gibbons GH, Dzau VJ, Izumo S (1993) J Clin Invest 92:2013
139. Tsao PS, Buitrago R, Chan JR, Cooke JP (1996) Circulation 94:1682
140. Ando J, Ohtsuka A, Korenaga R, Kawamura T, Kamiya A (1993) Biochem Biophys res
     Commun 190:716
141. Davies PF, Zilberberg J, Helmke BP (2003) Circ Res 2003 92:359
142. Kim BS, Nikolovski J, Bonaldo J, Mooney DJ (1999) Nat Biotech 17:979
143. Garvin J, Qui J, Maloney M, Banes AJ (2003) Tissue Eng 9:967
144. Sikavitsas VI, Bancroft GN, Holtorf HL, Lansen JA, Mikos AG (2003) Proc Natl Acad
     Sci USA 100:14 683
145. Mauck RL, Soltz MA, Wang CC, Wong DD, Chao PH, Valhmu WB (2000) J Biomech
     Eng 122:252
146. Sodian R, Lemke T, Fritsche C, Hoerstrup SP, Fu PF, Potapov EV, Hausmann H, Het-
     zer R (2002) Tissue Eng 8:863
147. Pioletti DP, Muller J, Rakotomanana LR, Corbeil J, Wild E (2003) J Biomech 36:131
148. Lee SH, Kim BS, Kim SH, Choi SW, Leong SI, Kwon IK, Kang SW, Nikolovski J,
     Mooney DJ, Han YK, Kim YH (2003) J Biomed Mater Res 66A:29
149. Brewster LJ, Trivedi SS (1984) IEEE-comput Graphics Applic 4:31
150. Vannier M, Marsh JL, Warren LO (1984) Radiology 150:179
151. Whimster WF, Cookson MJ, Salishbury JR (1995) Pathologica 87:279
152. Mihalopoulou E, Allein S, Luypaert R, Eisendrath H, Panayiotakis G (1999) Comput
     Methods Program Biomed 60:1
153. Berry E, Brown JM, Connell M, Craven CM, Efford ND, Radjenovic A, Smith MA
     (1997) Med Eng Phys 19:90
154. Colin A, Boire JY (1997) Comput Methods Program Biomed 53:87
155. Holck DE, Boyd EM, Mauffray RO (1999) Ophthalmology 106:1214
156. Petzold R, Zeilhofer HF, Kalender WA (1999) Comput Med Imaging Graphics 23:277

157. Hutmacher DW, Schantz JT, Zein I, Ng KW, Teoh SH, Tan KC (2001) J Biomed Mater Res 55:203
158. Zein I, Hutmacher DW, Tan KC, Teoh SH (2002) Biomaterials 23:1169
159. Hutmacher DW, Ng KW, Kaps C, Sittinger M, Klaring S (2003) Biomaterials 24:4445
160. Hollister SJ, Levy RA, Chu TM, Halloran JW, Feinberg SE (2000) Int J Oral Maxillofac Surg 29:67
161. Giordano RA, Wu BM, Borland SW, Cima LG, Sachs EM, Cima MJ (1996) J Biomater Sci Polym Ed 8:63
162. Griffith LG, Wu BM, Cima MJ, Chaignaud B, Vacanti JP (1997) Ann NY Acad Sci 831:382
163. Kim SS, Utsunomiya H, Koski JA, Wu BM, Cima MJ, Sohn J, Mukai K, Griffith LG, Vacanti JP (1998) Ann Surg 228:8
164. Sachlos E, Reis N, Ainsley C, Derby B, Czernuszka JT (2003) Biomaterials 24:1487
165. Lom B, Healy KE, Hockberger PE (1993) J Neurosci Methods 50:385
166. Singhvi R, Kumar A, Lopez GP, Stephanopoulos GN, Wang DI, Whiteside GM, Ingber DE (1994) Science 264:696
167. Clemence JF, Ranieri JP, Aebischer P, Sigrist H (1995) Bioconjug Chem 6:411
168. Chen G, Ito Y, Imanishi Y, Magnani A, Lamponi S, Barbucci R (1997) Bioconjug Chem 8:730
169. Chiu DT, Jeon NL, Huang S, Kane RS, Wargo CJ, Choi IS, Ingber DE, Whitesides GM (2000) Proc Natl Acad Sci USA 97:2408
170. Folch A, Toner M (1998) Biotechnol Prog 14:388
171. Delamarche E, Benard A, Schmid H, Michel B, Biebuyck H (1997) Science 276:779
172. Odde DJ, Renn MJ (2000) Biotechnol Bioeng 67:312

Adv Biochem Engin/Biotechnol (2006) 102: 139–159
DOI 10.1007/10_009
© Springer-Verlag Berlin Heidelberg 2005
Published online: 29 November 2005

Cellular to Tissue Informatics:
Approaches to Optimizing Cellular Function
of Engineered Tissue

Sachin Patil · Zheng Li · Christina Chan (✉)

Department of Chemical Engineering and Material Science, Michigan State University,
1257 EB, East Lansing, MI 48824, USA
krischan@egr.msu.edu

Abstract Tissue engineering is a rapidly expanding, multi-disciplinary field in biomedicine. It provides the ability to manipulate living cells and biomaterials for the purpose of restoring, maintaining, and enhancing tissue and organ function. Scientists have engineered various tissues in the body, from skin substitutes to artificial nerves to heart tissues, with varying degrees of success. Although the field of tissue engineering has come a long way since its first successful demonstration by Bisceglie in the 1930s, methods of coaxing them into functional tissues have been predominantly empirical to date. To successfully develop tissue-engineered organs, it is important to understand how to maintain the cells under conditions that maximize their ability to perform their physiological roles, regardless of their environment. In that context, a methodology that combines empirical data with mathematical and statistical techniques, such as metabolic engineering and cellular informatics, to systematically determine the optimal (1) type of cell to use, (2) scaffold properties and the corresponding processing conditions to achieve those properties, and (3) the required types and levels of environmental factors and the operating conditions needed in the bioreactor, will enable the design of viable and functional tissues tailored to the specific requirements of individual situations.

1
Tissue Engineering

Organ failure or loss of functional tissue is one of the more devastating and costly problems in medicine. Currently, organ transplantation is the only clinically proven effective treatment for patients with end-stage organ failure. Although organ and tissue transplantations are practiced widely, they are limited by a number of factors, the most important being the (increasing) shortage of donors. The latest data from the United Network for Organ Sharing (UNOS) indicate that over 25 000 organs were transplanted in 2002. However, the number of patients dying while on the waiting list has increased rapidly at a rate of approximately 11–13% per year. As of 31st August 2002, a staggering 80 856 patients were on the organ transplant waiting list [1]. Currently, the majority of organs donated are cadaveric, of which 92% are transplanted. Using marginal donors or domino transplantation expands the donor pool at the expense of increased risk of complications to the recipient. Once the organ is transplanted, the recipients are administered lifelong immunosuppressive medications, increasing their risks of infection, hypertension [2], hyperlipidemia [3] and cardiovascular diseases [4]. Due to these shortcomings, the field of tissue engineering has emerged as a potential alternative to organ transplantation, capable of addressing the shortage in organ availability [5]. Tissue engineering is an interdisciplinary field that integrates the recent developments in cellular and molecular biology with principles and methods of chemical and mechanical engineering in order to develop biological substitutes that can restore, maintain, or improve tissue function.

The approaches to tissue engineering can be broadly classified into two groups: genetic and cellular. The *genetic approach* is based on the idea of delivering DNA built into porous biomaterial carriers/scaffolds to the repair site. The biomaterial keeps the DNA in place until the endogenous cells arrive and are transfected/transduced to produce the therapeutic factor encoded by the DNA [6]. The advantage of this approach is its ability to induce a local tissue repair response rather than systemic changes in gene expression; however, the genetic approach to tissue repair addresses a fairly focused and specific set of situations. Alternatively, the *cellular technique* starts with a substrate or scaffold material, either natural or synthetic; cells are seeded onto the surfaces and exposed to various biochemical or mechanical stimuli. When the cells proliferate and migrate into the scaffold, they evolve into three-dimensional (3-D) tissue structures, which restore tissue function when implanted. As the tissue structure vascularizes and the scaffold dissolves, it eventually fuses into its surroundings and becomes a functional tissue. This approach has the advantage of a potentially broader applicability to a variety of situations [7].

Recent advances in tissue engineering have demonstrated the feasibility of using 3-D scaffolds to stimulate and guide regenerating tissue. The scaffolds are derived from a variety of natural and synthetic materials, from collagen, hydroxyapatite, Matrigel, alginate, to poly(α-hydroxy acids) polyesters, such as polyglycolic acid (PGA), polylactic acid (PLA), and copolymers of them (PLGA). Their material properties, such as their strength, rate of degradation, porosity, and microstructure, as well as their shapes and sizes, are more readily and reproducibly controlled in polymeric scaffolds [5]. Nonetheless, significant challenges still remain in developing suitable biomaterials that can mimic the mechanical properties of the tissues they intend to repair or replace, to minimize toxic degradation products, and to provide long-term biocompatibility [8, 9].

Although the development of designer biomaterials poses significant challenges, issues related to the cellular component, such as cell attachment, growth, differentiation and proliferation must be addressed concurrently in engineering viable tissue substitutes. The cells are the functional elements of the engineered tissue, whether growing and manipulating them in vitro for implantation or using them in an extracorporeal device. Methods to improve cell attachment are underway, such as modifying the polymer scaffold chemically or coating it with collagen. Similarly, approaches to aiding cell growth and proliferation, such as incorporating growth factors into the matrix, are being addressed [5, 10].

There are a variety of choices of potential cell sources, each with their own challenges. The use of xenogeneic cells (cells from a different species) for tissue repair offers a limitless supply of donor cells but harbors the potential risk of infecting humans with animal pathogens. Autologous (from the patient) and allogeneic (from a human donor) cells, such as skeletal muscle satellite cells, endothelial cells, and chondrocytes, have been used successfully to treat skin ulcers, other skin injuries, and burns [5]. This success has not been as easy to replicate for organs such as the liver and pancreas, partly because hepatocytes and pancreatic islet cells proliferate poorly in culture [11]. For this reason, embryonic stem (ES) cells are attractive and are currently being actively investigated, because they can proliferate in an undifferentiated state and subsequently induced to form various cell types [5, 12–15]. However, success can only be realized in this area if better markers to identify stem cells and their progeny and improved methods to isolate and proliferate these cells in vitro are found [16]. Additionally, further research is still needed to assess whether there is an immunological response to implanting cells derived from allogeneic donors [16].

Despite these challenges, tissue engineering has emerged as a potential alternative to tissue and organ transplantation. Over the past ten years, scientists have attempted to engineer tissues or organs in nearly every part of the body. The depth and breadth of the tissue-engineering field can be illustrated by the following recent examples. Eschenhagen et al. [17] have demonstrated the feasibility of engineering heart tissues (EHTs) from embryonic chick and neonatal rat cardiac myocytes. Cheng and Chen [18] have reported the successful fabrication of artificial nerves engineered from Schwann cells (SCs) isolated from peripheral nerve segments wrapped in polyglactin 910 fibers. Other tissues engineered include blood vessels [19], skin substitutes [20], bioartificial bladder [21], and tissue substitutes for bones [22], tendons [23], cartilage [24], ligaments [25], and kidney support system using encapsulated urothelial cells [26]. There has been recent clinical success with engineered blood vessels [27], extracorporeal liver assist devices [28, 29], and pancreatic islet implants to treat diabetes [30]. Thus, numerous organs and tissues have been engineered to varying degrees of success.

There is no doubt that tissue engineering has an enormous potential to make significant contributions to society over the next decade. Its success will largely depend on our ability to elucidate the complex interactions between genes, proteins, metabolites, and the environment that modulate cellular behavior. A more comprehensive methodology capable of extracting information on how all of these variables interact under a variety of conditions would enable a systematic approach to developing functional tissue constructs. To achieve this goal will necessitate an increasing reliance on computer-aided design and optimization methods to guide the manufacture of future engineered tissues.

2
Parameters for Tissue Engineering

To successfully develop tissue-engineered organs, it is crucial to understand (i) how to support and maintain cellular function and growth on biofunctional substrates [31, 32], (ii) how cells respond to external stimuli [10, 33, 34] and organize into structures that morphologically and functionally resemble tissues [35, 36], and (iii) how to maintain the cells under conditions that maximize their ability to perform their physiological roles, regardless of the environment, whether the cells are part of an extracorporeal system, such as the bioartificial liver assist device [37], or an implantable tissue-engineered device [11]. In either event, the cellular component of the device will eventually be exposed to the patient's plasma and must perform the various biological functions of the tissue, such as metabolic, anabolic, catabolic and other tissue-specific functions, and therefore must be able to survive and proliferate under a diverse set of environmental conditions: the patient's metabolic profile. Better insight into how these variables affect cell behavior and function would permit a more systematic approach to optimizing engineered tissues. In the examples that follow, we illustrate some of the variables (such as material composition, growth conditions) that have been studied for their affects on cellular functions (including cell spreading, protein synthesis) and the importance of optimizing these functions in the tissue construct.

Martz et al. [38] were the first to address the effect of the relative adhesion of cell-to-cell and of cell-to-substratum on cell culture morphology. Palecek et al. [39] further showed that variables such as the levels of substratum ligand and cell integrin expression and the integrin–ligand binding affinity affected short-term cell-substratum adhesiveness and cell migration. Bhatia et al. [40] used micropatterning techniques to further manipulate the cellular microenvironment to better control cell-to-cell interactions for potential tissue engineering applications.

Cell-to-substratum and cell-to-cell interactions play central roles in cell motility and spreading within the scaffold and organizing the cells into tissue formation [36, 41]. Ryan et al. [41] demonstrated that increasing the relative adhesive (cell-to-substratum) to cohesive (cell-to-cell) strength enhanced cell spreading into the scaffold. The adhesive strength of the cell-to-substratum was varied by changing the ratio of poly(desamino-tyrosyl-tyrosine ethyl ester) (DTE) to poly(ethylene glycol) (PEG) polymers incorporated in the substratum, while cell lines that expressed different types or levels of cadherins were used to vary the cohesive strength between the cells. This study clearly indicates the importance of these interactions and how they may contribute to the rational design of scaffold materials for engineering tissue constructs.

Optimizing growth conditions in the bioreactors depends upon the presence of key biochemical and mechanical factors that are essential to the development of successful engineered tissues [42]. For example, cells cultured on a monolayer are generally not nutrient-limited. In contrast, the supply of oxygen and soluble nutrients may be diffusion-limited in 3-D tissue cultures that are thicker than 100–200 microns [43]. The ability of the cells to proliferate and differentiate in these cultures declines with poor nutritional state. Therefore, studying the effects of variables such as O_2 tension, CO_2 glucose, ammonia, pH, and so on can help avoid inhibitory growth conditions [14, 42, 44], and supplements such as hormones and amino acids can enhance the long-term viability and function of the cells [45–49]. In addition to nutrient supply, effective removal of harmful metabolites is essential when optimizing the culture environment [50]. Thus, the cellular microenvironment is critical to in vitro cultivation of viable and functional tissues.

External stimuli, such as flow perfusion, pulsatile flow, and mechanical strains, have been shown to enhance the early differentiation and 3-D distribution of marrow stromal osteoblast cells [51], to increase hepatocyte survival and function after implantation [52] and to regulate the development of engineered smooth muscle tissue [34], respectively. Other variables, such as initial cell seeding concentration and the interval between seeding and addition of culture medium, have been studied for their effects on the seeding density within the tissue scaffold. The cell seeding density is evaluated to assess the degree of cell proliferation, aggregation and colonization within the scaffold [53, 54]. Determining an optimal cell seeding density within the scaffold is critical to the development of a dense tissue construct with a uniform cell distribution and high metabolic activity, which is especially important when engineering cardiac constructs [53]. This example illustrates the importance of characterizing the effects of environmental stimuli on the development of tissue structure and function.

The aforementioned studies demonstrate the breadth of variables that have been modulated or controlled to develop viable and functional tissue-engineered constructs. The approach taken in these studies has been to evaluate the effects of one or a handful of variables at a time. Alternatively, a more comprehensive methodology capable of extracting information on how all of these variables interact under a variety of conditions would enable a systematic approach to developing functional tissues. To achieve this goal will necessitate an increasing reliance on computer-aided design and optimization methods to guide the manufacture of future bioartificial tissues.

We propose that optimization approaches can be used to facilitate the rational design of engineered constructs capable of improving, restoring and maintaining tissue function. These approaches include elucidating the changes in the genetic and metabolic profiles of the cellular component in the construct relative to its environment (such as scaffold properties, nutrients, stimuli) and identifying the changes in the performance of the engineered construct as a function of these variables. This approach can help (i) to identify the environmental parameters and the corresponding levels of them necessary to achieve a targeted level of cellular growth or function, and (ii) to design potential solutions into the tissue construct, to avoid the development of conditions that may eventually lead to compromised long-term tissue function and survival of the recipient.

3
Metabolic Engineering Applied to Tissue Engineering

Understanding cellular processes is integral to our ability to manipulate cells, create novel and biocompatible scaffolds, and to identify key environmental variables that will lead to the development of functional tissue constructs. The ability to predict cellular responses as a function of genetic, metabolic, and environmental (including cell–cell and cell–substrate interactions) profiles is important, not only in enabling us to direct proliferation and differentiation pathways in vitro, but also in understanding how changes in these variables may influence their response in vivo. Thus, tissue engineering, which as Parenteau and Young put it, "began heavily focused on biomaterials, matrix biochemistry, and engineering, now adds a new level of advanced biological (cellular) considerations to the field going forward, distinguishing it from the more traditional medical device approach" [50].

The cell contains all of the genetic information required to reconstruct tissues and whole organs within itself, but its environment is critical in determining whether it will fulfill its fate. One thing is certain; if the right cells are placed in the right environment and given the right cues, they will develop into the intended tissue or organ [55]. This begs the question: what are the right environment and these right cues? Metabolic engineering has emerged as the technological and scientific discipline that can provide part of the answer. This multidisciplinary field integrates principles from chemical and mechanical engineering, material and computational sciences, biochemistry, and molecular biology. Metabolic engineering uses engineering principles of design and analysis to manipulate and improve cellular properties as a function of the cell's environment by modifying cellular pathways to achieve specific goals (chemical transformation and energy transduction), and so it has the potential to direct cells/tissue to more desirable states [56–58].

Before manipulating the cellular state, it is first necessary to generate a comprehensive map of the intracellular metabolic flow to elucidate the effects of the scaffold and the soluble factors in the environment on the metabolic state of the engineered tissue. A cellular state is defined as the metabolic phenotype taken on by the cells as a function of its environmental condition and genetic makeup. Thus far, most of the metabolic studies performed on engineered tissues have typically characterized the state of the cell/tissue by evaluating one or a handful of functional markers, such as P450 activity and urea for hepatocytes [59], and the insulin secretion profile as a function of glucose concentration for pancreatic islet cells [60]. However, measuring one or several markers provides insufficient information on the full range of functions/capabilities of the tissue that an otherwise more exhaustive evaluation would furnish. Thus, a more comprehensive method of characterizing the tissue state under a variety of environmental conditions will enable a more systematic approach to assessing the functional capabilities of the engineered tissue.

Characterizing the metabolic state of tissues requires that the reaction rates through the major intracellular pathways are determined. Studies have characterized the behavior of in vitro intermediary metabolism to better understand the regulation of metabolism of in vivo physiological and pathological states for mammalian systems [61, 62]. Several studies that have employed isotopic tracer methods to determine in vivo intracellular fluxes suffered from either low sensitivity, high cost, or require other tracer methods as well, thus increasing sensitivity to measurement errors [63, 64]. Methodologies relying on the use of stable labeled tracers such as ^{13}C-acetate and ^{13}C-lactate and appropriate mass balance models that relate the enrichment in the various isotope isomers (isotopomers) of specific intracellular metabolites, determined by NMR or gas chromatography–mass spectrometry, have been developed and thoroughly characterized [65]. One limitation of labeling methods is that they generally provide information about only one specific area of intracellular metabolism, such as the tricarboxylic acid cycle or gluconeogenesis, thus restricting their informative value and consequently insights into cellular metabolism of physiological and pathological states. These metabolic processes are coupled to other aspects of cellular metabolism using carbohydrate and amino acids via shared intracellular cofactors, such as NADH, ATP and NADPH, and metabolites such as acetyl-CoA, cysteine, and so on. An approach that can provide a comprehensive snapshot of the metabolic profile of the cells in the tissue construct as a function of its environment is metabolic flux analysis (MFA). The basis of this method is that the metabolic pathways have a well-defined stoichiometry relating reactants to products. Thus, the strength of this technique is its emphasis on the metabolic network rather than on individual reactions [56].

Metabolic engineering can be used to systematically identify pathways that can be manipulated to attain optimal tissue function. It can be used to alter cell characteristics such as growth, proliferation and substrate utilization [56]. Whatever the goal, whether it is to inhibit, create or enhance a cellular pathway or function, an analysis of the metabolic pathways provides a better understanding of the architecture as well as the potential of the metabolic network for direct manipulation of the cells for specific purposes. This permits a more directed approach to designing entire metabolic networks [66]. The quantitative analysis of metabolic networks and other metabolic engineering techniques has traditionally been applied to the optimization of cellular function, primarily for the purpose of enhancing the production of specialty chemicals from prokaryotic cells, yeasts, fungi, and to some limited extent, mammalian cells [67, 68]. For example, Sedlak et al. [69] succeeded in genetically cloning three xylose-metabolizing genes into *Saccharomyces cerevisiae* capable of fermenting xylose in addition to glucose to produce ethanol from different cellulosic biomass. Additionally, Sonderegger et al. [70] metabolically engineered a phosphoketolase pathway in *Saccharomyces cerevisiae*, by expressing phosphotransacetylase and acetaldehyde dehydrogenase in combination with the native phosphoketolase. This increased the rate of xylose fermentation by 40% and the corresponding yield of ethanol by 20%.

Less well recognized is the application of metabolic engineering tools to problems in physiology and medicine [56]. Isotopomer analysis of heart and liver metabolism has been used to quantify metabolic fluxes in the tricarboxylic acid (TCA) cycle and gluconeogenic pathway [37, 56, 63, 64, 71, 72]. For example, in vivo TCA cycle flux in the livers of rats infused with ethanol was measured with ^{13}C NMR spectroscopy [73]. Additionally, ^{13}C NMR spectroscopy was used to measure gluconeogenesis and pyruvate recycling in the rat liver [74]. Recently, ^{13}C NMR isotopomer analysis has been used successfully to study compartmentation of glycolysis and glycogenolysis in isolated rat heart [75]. More recently, nonisotopic MFA has been applied to quantify flux changes in perfused livers [76, 77] and cultured hepatocytes [78]. Further details on the specifics of MFA can be found in [79] and Lee et al. (in this volume).

4
Systems Approach to Optimizing Engineered Tissues

Typically, as mentioned in the previous section, the optimization process is performed empirically to obtain an optimal solution. However, an ideal optimization process should be able to determine systematically (1) the type of cell to use, (2) the scaffold properties and the corresponding processing conditions to achieve those properties, and (3) the optimal types and levels of environmental factors to supply and the operating conditions of the bioreactor to achieve the desired level of tissue function. In order to achieve these objectives, the optimization procedure must capture how the cells function and behave when confronted with changes in its environment. Given our current knowledge of biological systems, it is not yet feasible to develop a model from first principles that can describe the underlying mechanisms that govern cellular behavior as a function of its environment. Alternatively, we propose that one can capture this information from the experimental data itself, without defining a model a priori. This is based upon the supposition that if one can obtain a representative set of experimental results, then one could extract sufficient information from the data to capture how the cells behave as their environment changes. Appropriately with the advent of high-throughput technologies, genes, proteins, and metabolites for a given cellular or tissue state can now be easily quantified. Concomitantly, these technologies have spawned new, sophisticated and efficient bioinformatics approaches to analyzing and uncovering the complex biological system contained within the data. The data possesses the underlying structure and mechanisms of the biological system within it, providing a wealth of information from which regulatory and functional relationships may be unraveled. To illustrate this idea we will borrow an example from analytical chemistry. In spectroscopic analysis, full spectrum data (signal measurements) consist of absorbance values at different wavelengths, but only a subset of the spectral data is essential when building a calibration model to identify the chemical compounds being measured. In this case, the wavelengths are the underlying structure of the absorbance measurements (signal) and they determine the signals that are observed. Selecting an optimal subset of wavelengths maximizes the accuracy of the calibration model [80], and this optimal subset of wavelengths changes as the chemical species being measured changes. Similarly, we propose that one may infer the underlying structure of the gene regulatory network from the gene expression [81, 82] and from metabolic profiles [83]. By analogy, only a subset of these genes is relevant to a particular cellular function. Once the genes are identified, the gene regulatory network may be inferred with help from the literature. The subset of genes that are selected changes as the cellular function being predicted changes.

The goal of the bioinformatics approach is to be able to organize the collected data in a form that can uncover the relevant information hidden within the data. Various sophisticated mathematical techniques have been developed to handle this vast quantity of data with the objective of extracting this information and facilitating data interpretation. These modeling methods have been developed and in some cases adapted to process different types of data for a variety of situations. In general, the objectives of these approaches/models can be categorized into three optimization subgroups: feature selection, prediction, and diagnosis. 1) Feature selection is employed to extract a subset of representative variables by replacing the original variables with a smaller group of latent variables that can transform the original data into a simpler and more informative format. Latent variables represent linear or nonlinear combinations of the original set of variables. The model should provide a framework in which to evaluate patterns in the data and optimally identify "characteristic modes or features" within the data. These characteristic modes generally classify the data into groups, such as genes of similar regulation and function, or similar cellular state or biological phenotype. 2) Predicting a cellular response as a function of the environment and their genetic and metabolic profiles requires that the model be able to adequately capture this complexity from the data. The information contained within the data includes the linear and nonlinear relationships, the correlations between the external and internal variables and their regulation of the cellular function and behavior that is observed, and thus should be readily captured by the model. 3) Diagnosis requires a certain degree of knowledge of the cause–effect relationships among the variables, to properly identify the important factors that may be modulated to obtain optimal cellular or tissue function. Hence, the model should be able to extract this information from the data. Once these relationships are identified, they can be used to work out how to modulate the system variables (genes, pathways, and environmental factors) to obtain an optimal response in the target variable or cellular function. Therefore, it is necessary to identify and employ an appropriate bioinformatics technique for the specific optimization objectives.

5
Mathematical Approaches

Bioinformatics tools that can be employed to process the complex biological data include both linear approaches, such as principal component analysis (PCA), projection to latent structures (PLS), Fisher discriminate analysis (FDA) [84], multiblock PLS and PCA [85], and nonlinear approaches such as kernel PCA (KPCA), KPLS [86], support vector machine (SVM) [87], genetic algorithm [82], and graphical models such as Bayesian network analysis [81].

PCA seeks to find projections that maximize the variation in the data onto a lower dimensional projection space to facilitate data depiction. This method does not provide any discriminatory capabilities. Alternatively, FDA is a supervised pattern recognition method that uses linear projection analysis to transform the data to a reduced-dimensional space similar to PCA but is better at discriminating between groups of data. PCA merely seeks projections that maximize the variation in the data, thus permitting visualization of the sample distribution, while FDA maximizes the discrimination among the groups, thus permitting visualization of the sample separation between the different groups. FDA is a statistical technique for classifying sets of data into mutually exclusive and exhaustive groups a priori by defining a set of linear combinations of the independent variables. FDA characterizes the separation among groups by maximizing the ratio of between-group to within-group variability in order to maximize the discrimination and minimize misclassification errors in the reduced-dimensional space.

PLS is a technique that is amenable to interpreting large sets of highly correlated and often collinear data typical of experimental data, to provide diagnostic capabilities. PLS is a latent structure analysis, related to factor analysis, which attempts to simplify complex and diverse relationships that exist among observed variables by uncovering interrelationships that link the seemingly unrelated observed variables. PLS is a regression method with predictive modeling capabilities traditionally used in chemometrics [88] and process monitoring [85]. Typically in PLS, most of the relationships can be captured with the first few latent variables, while the remaining latent variables describe the random noise in the relationship, or the linear dependencies that do not strongly affect the underlying structure [89]. An advantage of this methodology is the insight it provides into the underlying structure of the data. The underlying structure captured by the latent variables may be a variable that was not measured, such as a transcription factor that affects the set of genes, or cannot be measured, as in transcription factor activities. With appropriate techniques, such as network component analysis (NCA), unobserved variables (transcription factor activities) may be inferred from observed variables (gene expression levels) [90, 91].

Multiblock PLS [16, 92], similar to PLS analysis, involves the determination of latent variables (linear combinations of the independent variables) to maximize the correlation between the independent and dependent variables. However, multiblock PLS, unlike PLS, allows one to efficiently investigate the structural relationship between multiple groups of independent variables and the dependent variable. Similarly, multiblock PCA is an extension of PCA in which the variables are divided into several blocks according to the nature of the variables, with principal components extracted from each block. For example, environmental variables may be contained in one block and metabolites in another, or cellular process such as TCA or urea cycle may be in one block and uptake of nutrients such as amino acids or glucose in another.

With the kernel technique, a kernel or function K can be defined such that for the sample data X and Z in attribute space (original data space), they can be mapped to feature space with a nonlinear mapping function φ,

$$K(X, Z) = \langle \varphi(x)\varphi(z) \rangle$$

The inner product of $\varphi(x)$ and $\varphi(z)$ can be implicitly calculated with the kernel function $K(X, Z)$ [86]. The advantage of the kernel method lies in the ability to map the data nonlinearly and implicitly to a feature space, where

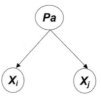

Fig. 1 A sample Bayesian network consisting of three nodes P_a, X_i, and X_j in which P_a is the parent node of X_i and X_j. Therefore X_i and X_j are conditionally independent of each other given the parent node P_a. P_a is the common cause of X_i and X_j, and X_i and X_j are the effects of P_a

linear methods such as PCA, PLS, FDA can be applied to train the data in this implicit feature space. This method permits the extraction of nonlinear features using linear methods and thus potentially side-steps computational problems typically associated with nonlinear mapping.

Bayesian network analysis has the ability to assess, with scoring metrics, causal relations based upon conditional probabilities, and so it permits hypothesis testing. Bayesian networks are directed acyclic graphs (DAG) whose nodes correspond to variables and whose arcs represent the dependencies between variables. The dependencies are determined by the conditional probabilities of each node x_i, given its parent node P_a, $Pr(x_i|P_a(x_i))$. A Bayesian network (i) assumes conditional independence, such that each node is independent of its nondescendants, given it parents; in other words, x_i and x_j are conditionally independent of each other given P_a (see Fig. 1), then

$$Pr(x_i|x_j, P_a(x_i)) = Pr(x_i|P_a(x_i)),$$

and (ii) consists of a set of variables $\{x_i\}$ defined by their joint distributions:

$$Pr(x_1, \cdots, x_n) = \prod_{i=1}^{N} Pr(x_i|P_a(x_i)).$$

Therefore, the entire network is a representation of the joint distribution defined by a set of random variables. This method was originally was applied to gene regulatory networks by [81, 93]. More recently, it has been used to reverse-engineer the structure of the metabolic network from experimental data and identify variables that are causally related to (or that regulate) a targeted cellular function [83].

6
Optimization of Cellular and Engineered Tissue Function

The engineered tissue contains three components: cells, scaffolds and bioreactors [94]. Cells are the functional elements in the engineered tissue, providing the basic biological function of the native tissue. Therefore, attaining optimal cellular function is of utmost importance. Cellular function can be optimized by modulating the expression of specific genes, as applied in metabolic control analysis (MCA) to control the amount of enzyme and in turn the rates of specific metabolic pathways [68]. MCA cannot predict a priori which gene and level of gene expression(s) are required to achieve a specific metabolic profile that will provide the level of cellular function desired. Furthermore, focusing on amplifying a specific enzyme may lead to metabolic imbalance and suboptimal productivity, which may however be overcome by engineering a regulatory circuit to control the gene expression in response to the intracellular metabolic state [95]. Alternatively, Schilling et al. [96] developed a framework that models the genotype–phenotype relationship. In their framework, a stoichiometric matrix of the entire metabolic network was reconstructed from the annotated genome of E. coli. The matrix was used in turn to obtain the metabolic flux distribution using flux balance analysis (FBA). The basic concept is that the metabolic genotype can be used to analyze, interpret and predict its metabolic phenotype. Then again, the annotated genome does not provide any information on post-translational modifications that may arise due to environmental affects, thereby altering the genotype–phenotype relationship. Therefore, it is necessary to acquire the changes in the metabolic and gene expression profiles that correspond to various experimental and environmental conditions in order to capture this information and thus better convey how genes, pathways, and environmental variables intertwine to regulate cellular function.

Developing a rational approach to optimizing cellular function involves an understanding of how the environment affects cellular function. The environment surrounding the cells modifies the interplay between the genetic and metabolic regulatory pathways and influences the cell's function and behavior. Therefore, characterizing the changes in the gene expression and metabolic profiles as a function of the environment will help to determine which variables (genes, metabolites and pathways) can be modified to achieve a desired function or level of cellular function.

With the help of microarray and metabolic engineering tools coupled with bioinformatics techniques, multiple genes and pathways can be simultaneously and readily identified and assessed for their role in modulating and optimizing cellular function. Microarray technology supplies the gene expression profile and MFA provides the metabolic flux distribution and in turn the metabolic profile of the cells as a function of their environment. Although these methods provide a comprehensive snapshot of the gene expression and metabolic profiles, a more sophisticated method is required to reveal the underlying structure in the data and the interplay of the variables involved in regulating the cellular response. Bioinformatics tools can be applied here, although the type of information obtained varies depending upon the method used.

For *feature selection*, FDA may be used to evaluate the underlying characteristic patterns of gene expression, metabolites or metabolic pathways as a function of alterations in the environment [97–99]. This method identifies the variables that contribute significantly to separating or "defining" a cellular state and making it distinctive from other states. Figure 2 illustrates an example of this method and the information it provides.

For *prediction*, FDA, PLS, KPLS, and multiblock PLS may be applied. PLS has been used to predict the expression level of a gene based upon the expressions of the other genes [100]. Similarly, it has predicted cellular function (the level of urea synthesized) based upon the metabolic profile and the environmental conditions [98]. PLS is appropriate for this type of situation, in other words a system containing more variables (genes) than the number of data sampled (replicates). Alternatively, FDA provides a different type of prediction. Given a map of the gene expression or metabolic profiles of the various samples in FDA space, it can identify which profile or set of samples is/are the closest match(es) to the unknown sample. For example, if one had images of two sets of sample populations, normal and cancerous, that mapped distinctly in FDA space, this technique can be used to evaluate the image of a new patient in order to determine which of the two groups his image belongs.

For *diagnosis*, PLS, multiblock PLS and Bayesian network analysis can be used to identify variables in the system that may be modulated to obtain a desired or optimal level of function. Figure 3 illustrates an example of the type of information multiblock PLS can provide. This method identifies which genes, pathways or variables are highly correlated to the target function being optimized. In other words, if a variable is highly and positively correlated, it suggests that increasing this variable will increase the level of the target function. Similarly, if they were negatively correlated, increasing that variable would decrease the target function. The type of diagnostic capabilities provided by multivariate statistical analysis (such as PLS or multiblock PLS), which is based on correlation analysis, is not always appropriate for determining causal effects. On the other hand, causal effects can be more readily assessed with conditional independence methods, such as Bayesian network analysis. The advantage of elucidating the causal relationship becomes apparent when faced with the situation depicted in Fig. 1. If increasing P_a increases both X_i and X_j, and if the goal is to decrease X_i, one may reason that it may be achieved by decreasing X_j. In this situation, X_i will not decrease, and in fact P_a must be altered instead to decrease X_i. Unlike multivariate statistical analysis, the Bayesian framework provides the ability to identify these causal relationships, which are important when identifying the *correct* variables to modulate for optimal function [83]. Alternatively, an advantage of multiblock PLS is that the model may be reversed to determine the optimal type and amounts of environmental factors required to attain optimal levels of cellular function [101].

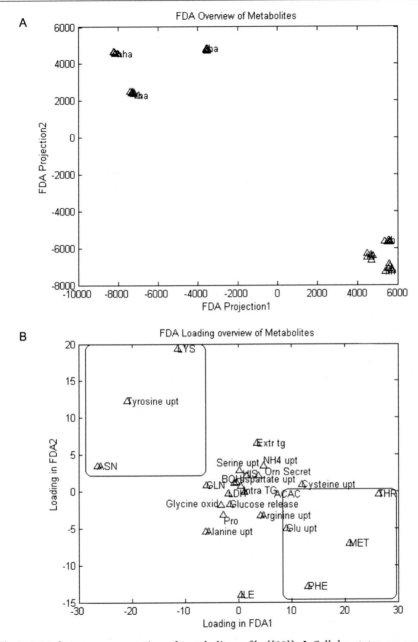

Fig. 2 FDA feature representation of metabolic profile [[98]]. **A** Cellular states are separated into six groups according to the FDA features extracted from the metabolic profile, and the group separation conforms with the experimental condition. **B** 2-D projection of the metabolite correlation. The figure shows the information about the metabolites that allows the groups to be separated (for example, those in the rectangles contribute to the separation of the amino acid supplemented samples)

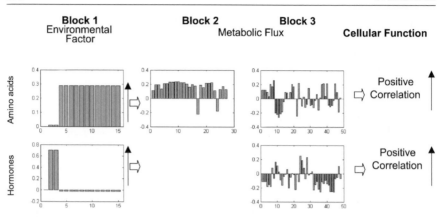

Fig. 3 The independent effects of an environmental block on cellular metabolism were identified based on the loadings in the environmental block (for example, the first LV in the MPLS model for the plasma supplementation period explained the effect of amino acids, while the second LV explained the effect of hormone supplementation). The loadings indicate that, for example, if we increased the fluxes that are positively correlated (with positive signs), the level of cellular function should increase. [[101]]

Currently, *feature selection* has been applied to select a subset of genes as "biomarkers" of the cellular or tissue state [82, 102]. With search methods such as sequential forward search, sequential forward floating search [102] and genetic algorithm [82], changes in the expression levels of the genes have been used to predict the cellular or tissue state. Ultimately, to fully capitalize upon high-throughput technologies, models that integrate information from the genetic up through the metabolic profiles need to be developed to capitalize upon a "systems approach" to optimizing cellular and tissue function.

Optimizing tissue engineered constructs is similar to optimizing cellular function, with the exception that the environmental variables now include the scaffold material and the operating conditions of the bioreactor in which the tissue construct is grown. Scaffolds are an essential part of the tissue engineering system, providing at the very least a support for cell attachment. When determining an appropriate biodegradable material for the scaffold, the chemical composition, breakdown products, mechanical properties and cell-scaffold interaction are important variables to evaluate and control [103]. In addition, identifying and controlling the scaffold processing variables is crucial to developing scaffolds with the optimal characteristics in terms of porosity, surface area, structural strength and specific 3-D dimensional shapes [104]. Bioreactors are used in tissue engineering to cultivate cells, produce the 3-D tissue scaffold and as an organ device support, such as in bioartificial liver (BAL) or kidney (BAK). It is essential to identify and control the parameters that determine the performance of a bioreactor in order to produce engineered tissues with the desired size, shape, biochemical composition and morphology. Generally, it involves parameters such as cell type, polymer scaffold, medium composition, exchange rates, bioreactor fluid dynamics, temperature, pH, PO_2, PCO_2, operating volume of the bioreactor, and many more. Ideally, adjusting these parameters can optimize the quality of an engineered tissue. Individual parameter optimization has been applied to identify the optimal rate of gas and nutrient exchange and bioreactor operating conditions in a chondrogenesis system (a tissue engineering system based on isolated cartilage cells (chondrocytes), biodegradable polymer scaffolds, and tissue culture bioreactors) [94]. Using the aforementioned optimization techniques, the quality of the engineered tissue can be expressed as a function of the cellular environment (the bioreactor, scaffold properties and operating conditions, as illustrated in Fig. 4). Methods such as PCA, PLS, MBPLS have been applied in chemical process control to identify the most significant factors that dictate product quality and the outliers that deviate from the normal or in-control states [85]. Likewise, these methods can be applied to determine how these various bioreactor parameters affect scaffold properties and tissue quality. The aforementioned optimization approaches can provide a means to identify "characteristic modes" such as operating parameters that provide a specific tissue state (*feature selection*), to quantitatively predict tissue function and quality based upon the operating parameters (*prediction*), and pinpoint key parameters that affect tissue function (*diagnosis*). Note that some variables such as cell type and polymer scaffold are ordinal variables that take discrete values. When modeling discrete variables, regression models such as logistic regression [105] and Bayesian network analysis [81, 83] can be applied to both discrete and continuous variables. In the case of the polymer scaffold, the chemical functional group (for example – COOH or – NH_3 groups) on the scaffold would be assigned "scores" that reflect the performance of the functional group on different surfaces or

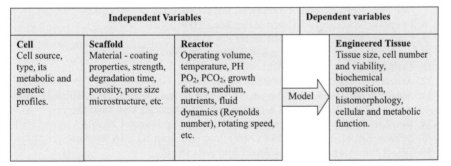

Independent Variables				Dependent variables
Cell Cell source, type, its metabolic and genetic profiles.	**Scaffold** Material - coating properties, strength, degradation time, porosity, pore size microstructure, etc.	**Reactor** Operating volume, temperature, PH PO_2, PCO_2, growth factors, medium, nutrients, fluid dynamics (Reynolds number), rotating speed, etc.	Model	**Engineered Tissue** Tissue size, cell number and viability, biochemical composition, histomorphology, cellular and metabolic function.

Fig. 4 Optimization of engineered tissue. A tissue engineering system is comprised of three components: cells, scaffolds and bioreactors. In order to systematically optimize the engineered tissue, a comprehensive characterization of the effect of each of the variables in the three components and their interactions on the function of the engineered tissue is needed

scaffolds; in other words the ability of the functional group to promote (for example) cell attachment or metabolic function.

Techniques such as PCA, FDA, PLS, MBPCA, MBPLS, KPCA, KPLS, SVM, genetic algorithm and Bayesian network analysis, can help determine the key parameters to monitor (or modulate) during the development of the tissue constructs. The parameters include the reactor operating conditions, scaffold properties, environmental, metabolic and genetic variables, which all play a role in regulating the cellular/tissue function. These techniques provide insight into the underlying mechanisms governing the tissue engineering system, offer a systematic approach to designing the tissue constructs, and may even identify potential solutions to ensure long-term viability and function.

Acknowledgements This work is supported in part by the National Science Foundation (BES 0222747 and BES 0331297) and the Whitaker Foundation. L.Z. is supported in part by the Center for Biological Modeling: Quantitative Biology Interdisciplinary Graduate Research Award at Michigan State University.

References

1. UNOS (2005) Organ distribution: allocation of livers (Policy 3.6). United Network of Organ Sharing, Richmond, VA (see http://www.unos.org/PoliciesandBylaws/policies/pdfs/policy_8.pdf, last accessed 6th October 2005)
2. Romero M, Parera A, Salcedo M, Abeytua R et al. (1999) Transplant Proc 31:2364
3. Gisbert C, Prieto M, Berenguer M et al. (1997) Liver Transpl Surg 3:416
4. Johnston SD, Morris JK, Cramb R, Gunson BK, Neuberger J (2002) Transplantation 73:901
5. Fuchs JR, Nasseri BA, Vacanti JP (2001) Ann Thorac Surg 72:577
6. Bonadio J (2002) Ann N Y Acad Sci 961:58
7. Fibrogen, Inc. (2005) Website. Fibrogen, Inc., San Francisco, CA (see http://www.fibrogen.com/tissue/, last accessed 6th October 2005)
8. Chapekar MS (1996) J Biomed Mater Res 33:199
9. Chapekar MS (2000) J Biomed Mater Res 53:617
10. Lee KY, Peters MC, Anderson KW, Mooney DJ (2000) Nature 408:998
11. Griffith LG, Naughton G (2002) Science 295:1009
12. Ariizumi T, Komazaki S, Asashima M, Malacinski GM (1996) Int J Dev Biol 40:715
13. Asashima M, Nakano H, Uchiyama H, Sugino H, Nakamura T, Eto Y, Ejima D, Nishimatsu S, Ueno N, Kinoshita K (1991) Proc Natl Acad Sci USA 88:6511
14. Hevehan D, Miller WM, Papoutsakis ET (2000) Exp Hematol 28:1016
15. Jackson KA, Mi T, Goodell MA (1999) Proc Natl Acad Sci USA 96:14482
16. Geladi P, Kowalski BP (1986) Anal Chim Acta 185:19
17. Eschenhagen T, Didie M, Munzel F, Schubert P, Schneiderbanger K, Zimmermann WH (2002) Basic Res Cardiol 97 Suppl 1:I146
18. Cheng B, Chen Z (2002) Chin J Traumatol 5:214
19. Weinberg CB, Bell E (1986) Science 231:397
20. Burke JF, Yannas IV, Quinby WC Jr, Bondoc CC, Jung WK (1981) Ann Surg 194:413
21. Oberpenning F, Meng J, Yoo JJ, Atala A (1999) Nat Biotechnol 17:149

22. Awad HA, Butler DL, Boivin GP, Smith FN, Malaviya P, Huibregtse B, Caplan AI (1999) Tissue Eng 5:267
23. Cao Y, Vacanti JP, Ma X, Paige KT, Upton J, Chowanski Z, Schloo B, Langer R, Vacanti CA (1994) Transplant Proc 26:3390
24. Stading M, Langer R (1999) Tissue Eng 5:241
25. Huang D, Chang TR, Aggarwal A, Lee RC, Ehrlich HP (1993) Ann Biomed Eng 21:289
26. Humes HD, Buffington DA, MacKay SM, Funke AJ, Weitzel WF (1999) Nat Biotechnol 17:451
27. Hibino N, Imai Y, Shin-oka T, Aoki M, Watanabe M, Kosaka Y, Matsumura G, Konuma T, Toyama S, Murata A, Naito Y, Miyake T (2002) Kyobu Geka 55:368
28. Chen SC, Mullon C, Kahaku E, Watanabe F, Hewitt W, Eguchi S, Middleton Y, Arkadopoulos N, Rozga J, Solomon B, Demetriou AA (1997) Ann N Y Acad Sci 831:350
29. Hu WS, Cerra FB, Nyberg SL, Scholz MT, Shatford RA (1997) US Patent 5 605 835
30. Lanza RP, Chick WL (1997) Ann N Y Acad Sci 831:323
31. Allen JW, Bhatia SN (2002) Tissue Eng 8:725
32. Langer R, Vacanti JP (1993) Science 260:920
33. Brieva TA, Moghe PV (2001) Biotechnol Bioeng 76:295
34. Kim BS, Nikolovski J, Bonadio J, Mooney DJ (1999) Nature 17:979
35. Alsberg E, Anderson KW, Albeiruti A, Rowley JA, Mooney DJ (2002) Proc Natl Acad Sci USA 99:12025
36. Lauffenburger DA, Griffith LG (2001) Proc Natl Acad Sci USA 98:4282
37. Yarmush DM, MacDonald AD, Foy BD, Berthiaume F, Tompkins RG, Yarmush ML (1999) J Burn Care Rehabil 20:292
38. Martz E, Phillips HM, Steinberg MS (1974) J Cell Sci 16:401
39. Palecek SP, Loftus JC, Ginsberg MH, Lauffenburger DA, Horwitz AF (1997) Nature (London) 385:537
40. Bhatia SN, Yarmush ML, Toner M (1997) J Biomed Mater Res 34:189
41. Ryan PL, Foty RA, Kohn J, Steinberg MS (2001) Proc Natl Acad Sci USA 98:4323
42. Obradovic B, Carrier RL, Vunjak-Novakovic G, Freed LE (1999) Biotechnol Bioeng 63:197
43. Ratcliffe A, Niklason LE (2002) Ann NY Acad Sci 961:210
44. Patel SD, Miller WM, Winter JN, Papoutsakis ET (2000) Cytotherapy 2:267
45. Evans ME, Jones DP, Zeigler TR (2003) J Nutr 133:3065
46. Franek F (1995) Biotech Bioeng 45:86
47. Furue M, Koshika S, Okamoto T, Asashima M (2000) In Vitro Cell Dev Biol Anim 36:287
48. Mastrangelo AJ, Betenbaugh MJ (1998) Trends Biotechnol 16:88
49. Washizu J, Berthiaume F, Chan C, Tompkins RG, Toner M, Yarmush ML (2000) J Surg Res 93:237
50. Minuth WW, Sittinger M, Kloth S (1998) Cell Tissue Res 291:1
51. Van Den Dolder J, Bancroft G, Sikavitsas V, Spauwen P, Jansen J, Mikos A (2003) J Biomed Mater Res 64A:235
52. Torok E, Pollok JM, Ma PX, Kaufmann PM, Dandri M, Petersen J, Burda MR, Kluth D, Perner F, Rogiers X (2001) Cells Tissues Organs 169:34
53. Dar A, Shachar M, Leor J, Cohen S (2002) Biotechnol Bioeng 80:305
54. Wiedmann-Al-Ahmad M, Gutwald R, Lauer G, Hubner U, Schmelzeisen R (2002) Biomaterials 23:3319
55. Godbey WT, Atala A (2002) Ann N Y Acad Sci 961:10
56. Jones JG, Carvalho RA, Franco B, Sherry AD, Malloy CR (1998) Anal Biochem 263:39

57. Lessard P (1996) Nat Biotechnol 14:1654
58. Yang YT, Bennett GN, San KY (1998) Elec J Biotech 1:134
59. Bergman RN, Van Citters GW, Mittelman SD, Dea MK, Hamilton-Wessler M, Kim SP, Ellmerer M (2001) J Investig Med 49:119
60. Kakela R, Kinnunen S, Kakela A, Hyvarinen H, Asikainen J (2001) J Toxicol Environ Health A 64:427
61. Crawford JM, Blum JJ (1983) Biochem J 212:595
62. Rabkin M, Blum JJ (1985) Biochem J 225:761
63. Katz J, Lee WN, Wals PA, Bergner EA (1989) J Biol Chem 264:12994
64. Katz J, Wals PA, Lee WN (1991) Proc Natl Acad Sci USA 88:2103
65. Yang D, Brunengraber H (2000) J Nutr 130:991S
66. Schilling CH, Schuster S, Palsson BO, Heinrich R (1999b) Biotechnol Prog 15:296
67. Nielsen J (1998) Biotechnol Bioeng 58:125
68. Stephanopoulous G, Aristidou AA, Nielsen J (1998) Metabolic engineering. Academic, San Diego, CA, p 203
69. Sedlak M, Ho NWY (2004) Appl Biochem Biotech 114:403
70. Sonderegger M, Schuemperil M, Sauer U (2004) Appl Environ Microbiol 70:2892
71. Jeffrey FM, Storey CJ, Sherry AD, Malloy CR (1996) Am J Physiol 271:E788
72. Large V, Brunengraber H, Odeon M, Beylot M (1997) Am J Physiol 272:E51
73. Jucker BM, Lee JY, Shulman RG (1998) J Biol Chem 273:12187
74. Jones JG, Naidoo R, Sherry AD, Jeffrey FMH, Cottam GL, Malloy CR (1997) FEBS Lett 412:131
75. Anousis N, Carvalho RA, Zhao P, Malloy CR, Sherry AD (2004) NMR Biomed 17:51
76. Arai K, Lee K, Berthiaume F, Tompkins RG, Yarmush ML (2001) Hepatology 34:360
77. Lee K, Berthiaume F, Stephanopoulos GN, Yarmush DM, Yarmush ML (2000) Metab Eng 2:312
78. Chan C, Berthiaume F, Lee K, Yarmush ML (2003) Biotechnol Bioeng 81:33
79. Vallino JJ, Stephanopoulos G (1993) Biotechnol Bioeng 41:633
80. Bangalore AS, Shaffer RE, Small GW (1996) Anal Chem 68:4200
81. Friedman N, Linial M, Nachman I, Pe'er D (2000) J Comp Bio 7:601
82. Li Z, Chan C (2004) J Biol Chem 279:27124
83. Li Z, Chan C (2004) FASEB J 18:746
84. Dillon WR, Goldstein M (1984) Multivariate analysis: Methods and applications. Wiley, New York, p 587
85. MacGregor JF, Jaeckle C, Kiparissides C, Koutoudi M (1994) Am Inst Chem Eng J 40:826
86. Schölkopf B, Smola AJ, Müller KR (1998) Neural Comput 10:1299
87. Cortes C, Vapnik V (1995) Mach Learn 20:273
88. Massart DL, Brereton RG, Dessy RE, Hopke PK, Spiegelman CH, Wegscheider W (1990) Chemometrics and intelligent laboratory systems, vol 1–5. Elsevier, Amsterdam
89. Baffi G, Martin EB, Morris AJ (1999) Comp Chem Eng 23:395
90. Li Z, Chan C (2004) Trends Biotechnol 22:381
91. Liao JC, Boscolo R, Yang YL, Tran LM, Sabatti C, Roychowdhury VP (2003) Proc Natl Acad Sci USA 100:15522
92. Nguyen DV, Rocke DM (2002) Bioinformatics 18:39
93. Pe'er D, Regev A, Elidan G, Friedman N (2001) Bioinformatics 17, Supl.1:s215
94. Freed LE, Vunjak-Novakovic G (2000) Principles of tissue engineering, 2nd edn. R.G. Academic, San Diego, CA, p 143
95. Farmer WR, Liao JC (2000) Nature Biotechnol 18:533

96. Schilling CH, Edwards JS, Palsson BO (1999a) Biotechnol Prog 15:288
97. Alter O, Brown PO, Botstein D (2000) Proc Natl Acad Sci USA 97:10101
98. Chan C, Hwang DH, Stephanopoulos GN, Yarmush ML, Stephanopoulos G (2003) Biotechnol Prog 19:580
99. Holter NS, Mitra M, Maritan A, Cieplak M, Banavar JR, Fedoroff NV (2000) Proc Natl Acad Sci USA 97:8409
100. Datta S (2001) Gene Expr 9:249
101. Hwang DH, Stephanopoulos G, Chan C (2004) Bioinformatics 20:487
102. Xiong M, Fang X, Zhao J (2001) Genome Res 11:1878
103. Pachence JM, Kohn J (1997) In: Lanza RP, Langer R, Chick WL (eds) Principles of tissue engineering, 1st edn. Academic, San Diego, CA, p 273
104. Thomoson RC, Yaszemski MJ, Mikos AG (1997) In: Lanza RP, Langer R, Chick WL (eds) Principles of tissue engineering, 1st edn. Academic, San Diego, CA, p 263
105. Hosmer D, Stanley L (1989) Applied logistic regression. Wiley, New York

Adv Biochem Engin/Biotechnol (2006) 102: 161–185
DOI 10.1007/b137205
© Springer-Verlag Berlin Heidelberg 2005
Published online: 25 October 2005

Review: Biodegradable Polymeric Scaffolds. Improvements in Bone Tissue Engineering through Controlled Drug Delivery

Theresa A. Holland · Antonios G. Mikos (✉)

Department of Bioengineering, Rice University, P.O. Box 1892, MS 142,
Houston, TX 77251-1892, USA
mikos@rice.edu

Abstract Recent advances in biology, medicine, and engineering have led to the discovery of new therapeutic agents and novel materials for the repair of large bone defects caused by trauma, congenital defects, or bone tumors. These repair strategies often utilize degradable polymeric scaffolds for the controlled localized delivery of bioactive molecules to stimulate bone ingrowth as the scaffold degrades. Polymer composition, hydrophobicity, crystallinity, and degradability will affect the rate of drug release from these scaffolds, as well as the rate of tissue ingrowth. Accordingly, this chapter examines the wide range of synthetic degradable polymers utilized for osteogenic drug delivery. Additionally, the therapeutic proteins involved in bone formation and in the stimulation of osteoblasts, osteoclasts, and progenitor cells are reviewed to direct attention to the many critical issues influencing effective scaffold design for bone repair.

Keywords Biodegradable polymers · Bone tissue engineering · Drug delivery

Abbreviations

bFGF	Basic fibroblastic growth factor
BMP	Bone morphogenetic protein
BSA	Bovine serum albumin
IGF-1	Insulin-like growth factor-1
OPF	Oligo(poly(ethylene glycol) fumarate)
PCL	Poly(ε-caprolactone)
PCPP-SA	Poly(carboxyphenoxy propane–sebacic acid)
PDGF	Platelet-derived growth factor
PEG	Poly(ethylene glycol)
PGA	Poly(glycolic acid)
PLA	Poly(D,L-lactic acid)
PLGA	Poly(lactic-co-glycolic acid)
PLLA	Poly(L-lactic acid)
PMMA	Poly(methyl methacrylate)
PPF	Poly(propylene fumarate)
rh	Recombinant human
ST-NH-PEG$_x$-PLA$_y$	N-succinimidyl tartrate monoamine poly(ethylene glycol)–poly(D,L-lactic acid)
TGF-β1	Transforming growth factor-β1
VEGF	Vascular endothelial growth factor

1
Introduction

The field of tissue engineering continues to revolutionize modern medicine by designing novel materials to restore tissue function. For the repair of large tissue defects, scientific efforts have demonstrated the utility of implanting tissue scaffolds, or solid substrates, to which cells can attach, allowing the ingrowth of new tissue [1]. Although the presence of a tissue scaffold is necessary in wound and tissue repair, repeated research has demonstrated that scaffolds alone often fail to provide a sufficient template to guide tissue regeneration [2–7]. Consequently, tissue engineering strategies must utilize biomaterials specifically designed to immobilize bioactive ligands, support cell transplantation, or deliver therapeutic molecules in order to achieve complete tissue repair [8–18]. While each of these approaches has been used to enhance healing in a number of clinical applications, bone defects are excellent candidates for local drug delivery strategies, since these defects often have access to the cells which growth factors target. Accordingly, the following discussion examines novel, degradable, polymeric scaffolds developed to locally deliver regulatory molecules for bone repair. A brief overview of the degenerative conditions affecting bone tissue, the cellular processes involved in bone formation and remodeling, and the regulatory molecules guiding these events is first provided so that readers clearly understand the physiological and biological challenges in bone tissue engineering. Subsequently, recent advances in drug delivery for bone repair are reviewed with an emphasis on polymeric scaffold design and the parameters affecting drug release.

2
Degenerative Bone Disorders

Trauma, congenital defects, and bone metastases frequently result in large bone deficiencies in both load bearing and non-load-bearing skeletal sites. Additionally, a number of other degenerative bone conditions are also in need of improved clinical therapies. Specifically, osteoporosis, Paget's disease, and rheumatoid arthritis lead to significant bone deterioration owing to the excessive proliferation and resorptive activity of osteoclasts [19]. Although hormone replacement therapies may help to inhibit bone resorption, oral ingestion of these pleotrophic agents has also been associated with an increased incidence of cancer and thromboembolic events [19, 20]. Furthermore, these treatment options, and the use of nondegradable bone cements to fill osseous voids, fail to stimulate osteoblasts to synthesize new bone. Although, poly(methyl methacrylate) (PMMA), a clinically used bone cement, has also been explored as a local delivery vehicle for chemotherapeutics, hormones, and antibiotics, observed release rates are generally too slow since this polymer is nondegradable [21–23]. Additionally, PMMA requires elevated curing temperatures (60 °C) which may engender further osteonecrosis [24, 25]. Accordingly, improved clinical strategies and materials for localized delivery of therapeutic molecules to bone defects are imperative.

3
Bone Formation

Bone tissue provides our bodies with an internal mechanical support system, while ensuring calcium homeostasis and housing the biological elements required for hematopoiesis, the process by which blood and immune cells are renewed [26]. This important tissue is formed from two distinct pathways. Intramembraneous bone formation, in which mesenchymal progenitor cells differentiate directly into osteoblasts, or bone-forming cells, leads to the development of the periosteal surfaces of the long bones, parts of the mandible and clavicle, and many cranial bones. Alternatively, the long bones and vertebrae are formed through endochondral bone formation in which mesenchymal progenitor cells differentiate first into chondrocytes. These cells deposit a cartilaginous template which is later mineralized and replaced by bone [26].

Upon fracture, bone is repaired by a process which recapitulates many of the events of both intramembraneous and endochondral bone formation [26]. Initially, after injury, a hematoma fills the defect site and may act as a source of signaling molecules to recruit reparative cells. An inflammatory response ensues as fibroblasts and macrophages replace this clot with an external callus, a fibrovascular tissue rich in collagen fibers. Within 7–10 days, intramembraneous bone, or hard callus, formation near the defect edges begins as osteoprogenitor cells of the periosteum differentiate into osteoblasts. Simultaneously, endochondral bone formation proceeds as additional progenitor cells differentiate into chondrocytes, which replace the external callus with cartilage. Finally, this cartilaginous, soft callus is mineralized and remodeled by the combined actions of both osteoclasts and osteoblasts [25, 26].

4
Bone Remodeling by Osteoclasts and Osteoblasts

Although bone remodeling follows fracture, this dynamic tissue is continually undergoing remodeling to maintain mechanical integrity and to respond to the body's changing demands [26, 27]. Remodeling results from the balance of two key processes, osteolysis, bone resorption by osteoclasts, and osteogenesis, bone formation by osteoblasts. Osteoclasts mature from macrophage precursor cells, and like many cells of the immune system, release numerous enzymes into the surrounding tissue. Osteolysis begins as the cell membrane of osteoclasts polarizes to secrete hydrochloric acid for dissolution of bone's inorganic, mineral component. Then, lysosomal protease and cathepsin K are mobilized to degrade the remaining organic component, mainly type I collagen fibers [27]. Osteoclast apoptosis marks the end of the resorptive phase of remodeling, and preosteoblastic cells are chemotactically recruited and differentiated into mature osteoblasts through numerous mitogens and growth factors. These cells fill the site with new bone matrix, which is completely mineralized within approximately two weeks [26, 28].

5
Regulatory Molecules

A host of bioactive proteins interact with the cell receptors on osteoblasts, osteoclasts, and progenitor cells to direct both new bone formation and bone remodeling. Specifically, these polypeptides may act as differentiation factors to guide progenitor cells toward a particular lineage or as progression factors to stimulate cell proliferation and extracellular matrix production [29–31]. However, many of these molecules function as both morphogens and growth factors [9]. In most cases, these soluble factors are synthesized and secreted by cells as precursor molecules which must be activated by binding to extracellular matrix components or by proteolytic cleavage [9, 29].

Bone morphogenetic protein-2 (BMP-2) is perhaps the most widely investigated agent in the field of bone tissue engineering. This molecule belongs to a group of proteins isolated from the inorganic component of bone by Marshall Urist, after researchers discovered that decalcified bone matrix could induce bone formation [32]. Advances in biochemistry led to the purification of at least 15 distinct molecules from this heterogeneous protein mixture, including BMP-2, BMP-4, BMP-7 (osteogenic protein-1), BMP-8 (osteogenic protein-2), and BMP-3 (osteogenin) [30]. These molecules mainly function as differentiation factors, guiding mesenchymal stem cells toward chondrogenic or osteoblastic lineage, and are osteoinductive agents capable of inducing bone formation ectopically [29, 30, 32, 33]. In fact, clinical trials are currently investigating the use of both BMP-2 and BMP-7 for bone repair [32].

BMPs actually belong to a broader superfamily of proteins, known as the transforming growth factor-β (TGF-β) family, and can share up to 50% sequence homology with TGF-β1 [29]. While the BMPs are osteoinductive agents, TGF-β1 appears to function as an osteoconductive agent, capable of inducing bone formation only in the vicinity of bone [29, 30]. Since this molecule is synthesized by osteoblasts and stored in bone matrix, bone serves as the body's largest reservoir of TGF-β1 [34]. This multifunctional protein has been shown to stimulate mesenchymal stem cell differentiation, enhance osteoblast proliferation, and inhibit osteoclast function [2, 26, 34]. However, at high doses, this pleotrophic agent is associated with inflammation, fibrosis, and scarring [35].

Additional signaling molecules involved in both bone repair and remodeling include basic fibroblastic growth factor (bFGF), platelet-derived growth factor (PDGF), and insulin-like growth factor-1 (IGF-1). Like TGF-β1, bFGF has been shown to modulate the functions of a number of cell types, such as osteoblasts, chondrocytes, fibroblasts, endothelial cells, and smooth muscle cells [26]. However, bFGF may be useful in bone tissue engineering since this protein can stimulate chondrocyte and osteoblast proliferation and enhance osteogenesis and angiogenesis [29, 30]. Like bFGF, PDGF and IGF-1 also encourage osteoblast expansion [26, 36]. Additionally, PDGF has been shown to act as a chemotractant and mitogen for mesenchymal progenitor cells [29, 31]. Both PDGF and IGF-1 stimulate the synthesis of collagen I and osteopontin, important organic components of the bone matrix [37].

6
Critical Issues in Drug Delivery System Design

Regardless of the bioactive protein or drug employed in delivery systems for tissue repair, engineers must ensure that these agents are released to the surrounding tissue within a therapeutic time frame and dosage. Thus, the resulting release profile should be optimized for each agent and clinical application. Since most morphogens and growth factors are relatively large proteins with precise three-dimensional conformations, care must be taken to ensure that the processing conditions used to fabricate drug delivery implants do not adversely affect the activity and half life of these agents. In particular, harsh loading conditions which promote protein aggregation or denaturation should be avoided [9, 38].

Since drug delivery systems for bone repair also serve as tissue scaffolds, these implants must conform to the design criteria used in the selection of all tissue engineering scaffolds. In addition to supporting cell adhesion, scaffolds should be biocompatible to prevent prolonged inflammation, as well as biodegradable to minimize the necessity of surgery [30, 39]. Ideally, scaffold materials should be metabolized by the body into acceptable degradation products at a rate which coincides with the rate of tissue ingrowth. Furthermore, drug-releasing scaffolds for bone repair must provide suitable mechanical support at the defect site and sufficient porosity to facilitate nutrient transport and cell infiltration [25]. Finally, implant materials should be chosen to allow for ease in sterilizing and processing scaffolds with a shape and a volume identical to those of the tissue defect [40].

A wide variety of natural materials have been used for the controlled delivery of ostegenic agents, including glycosaminoglycans, fibrin, alginate, gelatin, and collagen [41–47]. Initially, demineralized bone matrix, obtained from the cortical bone of various animal sources, was widely investigated in bone repair, since this mixture of collagenous proteins is a natural carrier of BMPs [48, 49]. However, these materials often fail to possess the mechanical properties and resorption rates necessary for load-bearing applications [50]. Alternatively, the physical and chemical parameters of polymeric devices can be easily and reproducibly tailored for a given application [24, 40]. Furthermore, implantation of delivery scaffolds based on synthetic polymers circumvents concerns regarding immunogenic reactions and disease transmission from materials derived from allogenic or xenogenic tissues. The following discussion examines many biodegradable, polymeric drug delivery systems for bone tissue repair. As illustrated by in vitro and in vivo research, polymer composition, hydrophobicity, crystallinity, and degradability, as well as the method of drug loading, are among the many properties affecting drug release and tissue formation.

7
Polymeric Scaffolds for Bone Drug Delivery

Perhaps one of the most influential material parameters dictating drug release centers on carrier degradability. In general, biodegradable polymers possess hydrolytically unstable linkages in their backbone, such as ester, ether–ester, anhydride, or amide functional groups [40, 51, 52]. Often scientists classify the mechanism by which hydrolysis proceeds as either bulk degradation or surface erosion. In the case of bulk degradation, the rate at which water penetrates an implant exceeds the rate at which the polymer is converted into water-soluble fragments, resulting in material deterioration throughout the device. Surface erosion refers to the opposing situation, when water penetration occurs more slowly than the rate of polymer solubilization, allowing bulk integrity to be maintained as surfaces exposed to water erode [40, 53, 54].

Systems which undergo bulk degradation often display first-order diffusion-controlled release kinetics and may be too hydrophilic for drugs that are highly unstable in an aqueous environment. Release from surface-eroding polymers typically proceeds at zero-order rates controlled by the rate of surface degradation. Accordingly, these materials tend to be more hydrophobic and may better preserve the activity of molecules within the polymer matrix [55, 56]. Polyesters, polyether-esters, and polyester-amides are generally classified as bulk-degrading polymers, while polyanhydrides and polyorthoesters degrade primarily by surface erosion [7, 40, 55, 56]. However, engineers should be aware that changes to the microarchitecture and surface properties of drug-release systems, stemming from mechanical loading, fibrosis, or vascularization of implants, may alter both polymer degradation and the release and activity of incorporated proteins [9, 30]. Additionally, since many drug delivery scaffolds are fabricated from two or more polymers, hydrolysis often proceeds through both bulk and surface degradation.

7.1
Poly(lactic acid)

The most widely investigated polymeric drug carriers have arisen from a group of degradable aliphatic polyesters approved by the FDA for use in other clinical applications, including surgical sutures, pins, clips, and staples [9, 54, 55]. Early drug delivery research sought to extend the use of these polymers for the sustained local release of antibiotics as an alternative to treating postoperative osteomyelitis by lengthy oral or intravenous drug regimes [57–59]. For instance, the polyester poly(lactic acid) was explored as a coating for implants releasing gentamicin and allowed for sustained release of this antibiotic at the minimum inhibitory concentration toward the bacteria *Staphylococcus aureus* for approximately four weeks [60]. Given these successes, engineers have extended the use of these polyesters for the design of more complicated systems for tissue regeneration.

The structure of poly(lactic acid) is shown in Table 1 along with numerous polymers investigated as carriers of therapeutic molecules for bone repair. Poly(lactic acid) is formed by ring-opening polymerization of lactide, the dimerization product of lactic acid. Two optical isomers of lactic

Table 1 Biodegradable polymers utilized in osteogenic drug delivery

Polymer	Chemical structure	Bone engineering research
Poly(lactic acid)		[6, 67–69] [57, 58, 60]
Poly(lactic-*co*-glycolic acid)		[3, 4, 31, 50, 61–63, 66, 71, 72, 74, 76–80]
Poly(ε-caprolactone)		[82–84]
Polyglyconate		[40]
Poly(lactic acid)-poly(ethylene glycol)		[10, 87–90, 94, 96, 97]

Table 1 (*continued*)

Polymer	Chemical structure	Bone engineering research
Poly(ethylene glycol)diacrylate		[95]
Oligo(poly(eythlene glycol) fumarate)		[16, 18, 98–101, 118]
Poly(carboxyphenoxy propane-sebacic acid)		[5, 102–106]
Poly(propylene fumarate)		[81, 107–112, 120, 126]

acid exist, corresponding to L-lactide or D-lactide. Poly(L-lactic acid) (PLLA), formed from the naturally occurring isomer, is a semicrystalline polymer with a relatively high melting point (178 °C) and glass-transition temperature (65 °C). Accordingly, these properties impart high tensile strength and extended degradation times (3–5 years) to PLLA scaffolds. Poly(D,L-lactic acid) (PLA), polymerized from a blend of D-lactide and L-lactide, is amorphous, and thus possesses considerably lower melting point (60 °C), tensile strength, and degradation times (12–16 months) than PLLA [40, 55]. However, since the degradation of PLLA yields the naturally occurring stereoisomer of lactic acid, a normal intermediate of carbohydrate metabolism, this form of poly(lactic acid) is generally preferred [54, 55].

Early growth factor release systems based on PLLA and other polyesters utilized simple adsorptive techniques to surface-coat polymeric particulates [61–66]. Nonuniform particulates were obtained by such means as the shredding of solid polymer rods [63]. Often the protein of interest was dissolved in a solution containing additional agents, such as collagen, serum, chitosan, or carboxymethylcellulose, to promote sorption to the polymer. However, these particulates were not able to function as true tissue scaffolds owing to their lack of mechanical support. In fact, to maintain particles at a defect site, gelatin or collagen capsules were sometimes employed [63, 64], making it difficult to clearly assess device efficiency and to correlate in vivo and in vitro behavior. Such delivery systems were primarily investigated in non-loading-bearing models and often displayed inadequate bone regeneration.

Despite their limitations, these early release systems assisted in elucidating valuable information regarding various biological–material interactions. For instance, systematic investigations with polymeric particulates helped to optimize the PLLA molecular weight range appropriate for bone implants [67]. Specifically, PLLA particulates (100 mg) of various molecular weights were wetted with 4 mg semipurified BMP and then implanted in the dorsal muscles of mice. No bone formation was visible after three weeks with particulates of high molecular weight (greater than 3300 Da), since the slow degradation rate of these polymers restricted tissue ingrowth. Tissue necrosis was observed with particulates of extremely low molecular weight (160 Da), owing to the high acidity and rapid degradation of these formulations. However, an intermediate PLLA molecular weight (650 Da) resulted in limited bone formation [67]. Additional studies compared bone regeneration from polymer-based systems to the previous standard carriers, collagen or demineralized bone, and demonstrated that PLLA and other polyesters did not adversely affect the activity of released growth factors [50, 62].

With the development of more advanced scaffold fabrication techniques, such as solvent-casting, gas-foaming, and emulsion freeze-drying, PLLA was formulated into three-dimensional scaffolds for drug delivery to loading-bearing defects [38]. For instance, 50 μg recombinant BMP-2 (rhBMP-2) was reconstituted in a collagen solution and then adsorbed to the surface of prefabricated PLLA disks before implantation in critical-size rat calvarial defects. After four weeks, defects treated with these delivery systems demonstrated enhanced bone formation by radiomorphometric and histomorphometric analysis when compared with PLLA disks seeded with osteoprogenitor cells and unloaded PLLA controls [6].

However, more precise control over protein-release kinetics, and thus substantially lower, yet still therapeutic, drug loadings, were often achieved by incorporating the desired protein directly into the polymer network [68, 69]. For example, using an air-drying phase-inversion process, PLLA was mixed with 200 ng PDGF and fabricated into porous membranes. In vitro release experiments demonstrated sustained PDGF release over the course of 28 days. Release rates could be increased through dual loading of both PDGF and bovine serum albumin (BSA) into these matrices. Furthermore, bone formation within critical-size rat calvarial defects was achieved using these delivery systems within approximately two weeks after implantation, demonstrating maintenance of protein activity with this scaffold fabrication technique [68].

7.2
Poly(lactic-co-glycolic acid) Copolymers

Additional efforts to achieve further control over protein release led to the copolymerization of lactide and glycolide to form poly(lactic-co-glycolic acid) (PLGA). Pure poly(glycolic acid) (PGA) was used to develop DEXON, the first commercially available synthetic absorbable suture [55]. Like PLLA, PGA is formed by ring-opening polymerization of a cyclic dimer and results in a semicrystalline network [40]. However, copolymerization of lactide with glycolide disrupts the crystallinity of these monomers, leading to an amorphous network with a rate of degradation and protein release dictated by the monomer ratio [54]. PLGA with a lactide-to-glycolide ratio of 7 : 3 is used commercially to produce surgical staples marketed as Lactomer [55]. Solvent-casting/porogen-leaching techniques have been utilized to fabricate porous PLGA foams which support osteoblast and mesenchymal stem cell attachment and proliferation [70–72].

To deliver therapeutic agents from PLGA-based materials, a number of scaffold processing techniques have been employed. Emulsion freeze-drying methods have been used to fabricate and model release of active rhBMP-2 from scaffolds of various pore sizes [73, 74]. Innovative methods for encapsulating proteins within PLGA microspheres have also been developed [4, 75]. The most popular of these methods, a double-emulsion-solvent-extraction technique, has been shown successfully to entrap rhBMP-2, TGF-β1, and IGF-1 within polymeric microparticles without significant loss in protein activity [76–78]. Protein-release kinetics from PLGA microparticles can be controlled by altering the loading of additional components, including poly(ethylene glycol) (PEG), BSA, and gelatin [78, 79].

Utilizing these and other processing techniques for the fabrication of both PLGA microparticles and scaffolds, researchers now have the ability to create composite materials with precise release profiles [80, 81]. For instance, delivery of multiple growth factors at specific rates is now possible. Slow release of PDGF (approximately 0.1 pmol/day) and fast release of vascular endothelial growth factor (VEGF) (1.7 pmol/day) was achieved using a novel scaffold design. PDGF was first encapsulated within PLGA microparticles, while VEGF was mixed PLGA particulates. The microparticles and particulates were then combined and processed into porous foams using a gas-foaming technique. Four weeks after implantation in the hind limbs of mice, these dual-release systems demonstrated enhanced vasculature when compared with systems releasing only PDGF or VEGF [80]. More importantly, similar polymeric devices will provide researchers with a tool to assess how various growth factors affect tissue repair. Additionally, these systems can be utilized to optimize the therapeutic release profiles of particular proteins in bone healing.

7.3
Other Polyesters

Unfortunately, the mechanical properties of scaffolds based on PLGA are well below the mechanical properties of human trabecular bone [70]. Thus, researchers have investigated several other polyesters for bone engineering applications. For instance, porous scaffolds fabricated by solvent-casting/porogen-leaching and based on the polyester poly(ε-caprolactone) (PCL) demonstrated significantly higher tensile strength and Young's modulus than PLGA scaffolds [82]. However, like PLLA, PCL is a semicrystalline polymer with a degradation time of approximately two years [40]. To speed degradation, copolymers of ε-caprolactone and DL-lactide have been synthesized. However, when implanted in femoral defects in rats, this material still remained after one year and appeared to retard bone formation when compared to untreated defects [83]. Other researchers have taken advantage of the lengthy degradation time of highly crystalline polymers to protect bone grafts from displacement and rapid resorption [84, 85]. However, their slow degradation has limited the extension of these materials to drug delivery.

Following a similar strategy to tailor both polymer strength and degradability to tissue engineering applications, additional research groups have synthesized PGA-based copolymers. In particular, polyglyconate, a copolymer of glycolide and trimethylene carbonate, is utilized in surgical sutures, tacks, and screws. Copolymers with a 2 : 1 glycolide-to-trimethylene carbonate ratio demonstrate increased flexibility and faster degradation times (seven months) than pure PGA [40]. Similar trends are reported when glycolide is polymerized with both trimethylene carbonate and/or p-dioxanone [40, 86]. Yet, like lactide-based copolymers, these materials have not been widely used in drug delivery.

7.4
Poly(lactic acid)-Poly(ethylene glycol) Block Copolymers

One group of copolymers which has been investigated as a protein carrier is the family of PLA_y-PEG_x block copolymers. These copolymers consist of alternating segments of PLA and PEG, whose respective molecular weights y and x are dictated by the polymerization reaction. The hydrophilic nature of the repeating PEG unit helps to neutralize the acidity of low molecular weight PLA segments, while modulating the degradation rate [87]. When loaded with rhBMP-2, block copolymers with a 7 : 3 PLA-to-PEG ratio demonstrate promising bone formation. More specifically, 10 µg rhBMP-2 was incorporated into $PLA_{6500}PEG_{3000}$ devices, and then surgically implanted into the back muscles of mice. After approximately three weeks, only 21% of the polymer remained, and bone formation with osseous trabeculae was apparent. However, BMP-2 release from implants with a higher PLA-to-PEG ratio did not result in osteoinduction owing to the extensive amount of polymer remaining (94–98%) after three weeks [88]. The optimal total block copolymer molecular weight for BMP-releasing implants with a PLA-to-PEG ratio of 7 : 3 was found to be 6400 Da, allowing for complete in vivo degradation by three weeks [89].

Additional research demonstrated that the degradation rate of these block copolymers could be further tailored by introducing a random linkage of p-dioxane. Furthermore, these materials, when loaded with 0.5–10 µg rhBMP-2, were able to illicit bone formation intramuscularly in rats [90]. Accordingly, these studies demonstrate that as the material properties of a delivery system are improved, the minimal effective dose of a particular drug or growth factor is often substantially reduced. However, further experimental investigations are necessary to understand the interplay between in vivo material degradation, protein-release kinetics, and tissue formation.

While such investigations are ongoing, a novel means of immobilizing proteins within scaffolds was developed by modifying these copolymers. Specifically, N-succinimidyl tartrate monoamine PEG-PLA (ST-NH-PEG_x-PLA_y), an amine-reactive polymer, was synthesized using innovative chemical techniques [91]. The amine group of this polymer was shown to facilitate covalent attachment of model proteins in both solution (insulin and somatostatin) and the solid phase (trypsin). Similar results were found with a novel thiol-reactive polymer, synthesized by attaching N-succinimidyl 3-maleinimido propionate to monoamine PEG-PLA [91]. Although the bioactivity of the attached proteins has not yet been confirmed in this system, in vitro and in vivo studies with similar protein tethering methods suggest that these methods should not alter protein activity [92, 93].

Taking this technology even further, an inventive means of fabricating scaffolds with interconnected pore networks was developed through incorporation of lipid microparticles during ST-NH-PEG_x-PLA_y precipitation into n-

hexane. This method avoids an aqueous environment, preserving the amine group from hydrolysis, and thus permitting the covalent attachment of proteins. Upon polymer precipitation into three-dimensional structures, lipid microparticles are subsequently extracted by melting to yield porous scaffolds [94]. Such creative techniques will undoubtedly revolutionize tissue engineering, allowing for the creation of truly biomimetic structures.

7.5
Additional Poly(ethylene glycol)-Based Materials

Toward the goal of minimizing implantation surgeries, researchers have also been developing novel injectable polymers from the delivery of bioactive molecules. Many of these polymers have a repeating PEG unit within their backbone, since the hydrophilicity of PEG often imparts water solubility. Upon thermal or photoinitiated reactions, these polymers are converted from their soluble state into crosslinked hydrogels. For instance, macromers of PEG with acrylated end groups have been photo-cross-linked into hydrogels and utilized for protein delivery [95]. Likewise, macromers of PLA-PEG-PLA with acrylated end groups have been examined as osteogenic protein or cell delivery vehicles [10, 96, 97]. However, these systems have mainly been studied in vitro or in subcutaneous implantations.

Another PEG-based macromer, oligo(PEG fumarate) (OPF) has been extensively examined in vitro [16, 18, 98] and utilized in vivo for the repair of both bone and soft tissue defects by modification with bioactive peptides [99]. The ester linkage in the backbone of this macromer facilitates hydrolytic degradation, while the double bond facilitates crosslinking through thermal initiation. Thus, OPF can be crosslinked into degradable hydrogels at physiological temperatures [100]. Furthermore, TGF-β1 release from OPF hydrogels can be tailored by altering the swelling ratio and mesh size of these networks [101]. However, since all of these PEG-based materials form water-absorbent gels, their utility will be limited to soft tissue defects and non-load-bearing bone defects. Additionally, this characteristic prevents their use as a carrier for drugs which are highly unstable in an aqueous environment.

7.6
Polyanhydrides

Alternative carriers for hydrolytically unstable molecules are often based upon more hydrophobic polymers like polyanhydrides. These polymers are synthesized by melt polycondensation. Typically, a diacid monomer is reacted with excess acetic anhydride, yielding an anhydride oligomer, which can then be polymerized under vacuum [56]. Homopolyanhydrides of aliphatic or aromatic diacid monomers generally possess some degree of crystallinity. As discussed previously, copolymers show a decrease in crystallinity owing to the presence of other units in the polymer chain, and the degree of monomer hydrophobicity will dictate the polymer's degradation time [55, 56].

The most widely investigated polyanhydride is a copolymer of sebacic acid and 1,3-bis(p-carboxyphenoxy)propane, know as poly(carboxyphenoxy propane-sebacic acid) (PCPP-SA) [40, 55]. Copolymers with high sebacic acid contents demonstrate relatively short degradation times owing to the relative hydrophilicity of this monomer [102, 103]. Drug release from PCPP-SA and other surface-eroding polymers has been shown to directly coincide with the rate of degradation [36, 51, 55, 56]. Like polyesters, early polyanhydride release systems were examined for the localized delivery of antibiotics to treat osteomyelitis [104]. However, PCCP-SA and other polyanhydrides have been successfully fabricated into microparticles and scaffolds for the controlled delivery of a number of different bioactive molecules.

For instance, a hot melt microencapsulation procedure was developed to entrap a model drug (insulin) within PCPP-SA microparticles [105]. The biological activity of insulin released from these microparticles was later verified. In particular, insulin-incorporated microspheres injected into diabetic rats resulted in normoglycemia for a period of approximately five days [102]. Scaffolds have also been fabricated for the controlled delivery of active proteins. Using compression molding, a mixture of osteogenic proteins and polymer was fabricated into polyanhydride scaffolds based on several different macromers [5]. Protein release from these systems was shown to induce osteogenesis intramuscularly in mice. However, osteoinduction did not result in animals receiving an injection of the same dose of drug, due to protein solubility, and thus, distribution with in the body [5].

In similarity to the challenges facing polyester-based systems, considerable work is necessary to enhance the performance of polyanhydride scaffolds so that these materials provide the appropriate mechanical properties to ensure tissue repair. Accordingly, researchers are investigating the use of dimethacrylated anhydride monomers which can be crosslinked into networks with improved mechanical properties. In particular, dimethacrylated sebacic acid and 1,3-bis(p-carboxyphenoxy)hexane were photo-crosslinked into networks with respective compressive strengths of 34 ± 4 and 39 ± 11 MPa. While these values are comparable to the compressive strength of trabecular bone (5–10 MPa), further improvement is necessary to reach the strength of cortical bone (130–220 MPa) [106].

7.7
Poly(propylene fumarate)

A material that has been shown to form networks with compressive strength on the order of cortical bone is the linear unsaturated polyester poly(propylene fumarate) (PPF) [107]. The double bond in the backbone of PPF allows this viscous polymer to be crosslinked into solid structures via thermal initiation or photoinitiation [76, 108]. By altering the molecular weight, adding PPF-diacrylate, and varying the crosslinking reaction, the compressive strength of PPF-based networks can be tailored within the range of 31 ± 13 to 129 ± 17 MPa [107]. Furthermore, PPF scaffolds with an interconnected pore network can be easily fabricated by incorporating a porogen during the crosslinking reaction [109].

These scaffolds have been shown to be both biodegradable and biocompatible in vivo [110]. Their use as delivery vehicles has also been explored. Preliminary work investigating the repair of rabbit cranial defects demonstrated significant bone formation when defects were treated with PPF scaffolds to which rhTGF-β1 was adsorbed [111]. More sophisticated systems for the controlled delivery of osteogenic agents have also been developed using PPF. In particular, microparticles encapsulating the therapeutic peptide TP508 have successfully been integrated into both the pores and the polymer network of PPF scaffolds for the sustained release of this molecule [81]. These systems were then implanted into critical-size rabbit femoral defects to evaluate how the kinetics of TP508 release affects bone repair. Complete bone regeneration was observed when TP508-loaded microparticles were incorporated into the pores of these scaffolds, indicating that a burst release of this molecule promotes bone formation. However, slow release of TP508 from microparticles within the polymer network demonstrated considerably less bone formation [112]. Thus, scaffolds based on this novel polymer not only impart improved mechanical properties, but also provide a means of controlled protein delivery to systematically evaluate the therapeutic time course of individual proteins.

8
Future Directions

As illustrated in the preceding discussion, researchers have developed many novel materials for the delivery of osteogenic agents. While initial release systems focused on the use of simple aliphatic polyesters, the shortcomings of these materials have led researchers to synthesize various polyester copolymers, as well as alternative polymers like PCPP-SA and PPF. Still other researchers have developed injectable polymers to minimize the necessity of surgery [8, 88, 100, 113, 114]. Meanwhile, scientists outside the arena of polymer chemistry have also contributed significantly to the field of bone tissue engineering.

For instance, biologists have not only isolated a number of therapeutic proteins for bone repair, but have also investigated the affinity of these agents to extracellular matrix components [42, 45, 115–117]. Accordingly, new composite systems are now incorporating some of these components into synthetic polymeric scaffolds to allow growth factor release to be catalyzed by natural processes [118]. Additionally, mechanical engineers have shown that hydroxyapaptite and β-tricalcium phosphate can be incorporated into synthetic polymeric scaffolds to improve the mechanical properties of these networks [82, 86, 119, 120]. Scientists in the field of nanoscience are also contributing to the field of tissue engineering, demonstrating that nanoreinforcement of polymers may assist in building stronger scaffolds [121, 122]. Methods for encapsulating proteins within nanoparticles have also arisen [123–125]. Finally, advances in computer-aided design have led to the fabrication of scaffolds of precise three-dimensional shape and volume [1, 126]. Accordingly, researchers now have the material and processing tools to build more effective polymeric scaffolds for drug delivery. However, to develop highly efficient release systems, engineers must now work closely with biologists to study how the kinetics of protein release from these systems affects bone repair.

9
Conclusions

A host of bioactive proteins have been isolated from bone tissue and shown to influence the cellular and molecular events involved in both bone formation and bone remodeling. Likewise, numerous degradable polymers have been developed as carriers of these molecules. These polymers include the early polyesters, such as PLLA and PLGA, as well as a later generation of injectable materials based on PEG, and more hydrophobic materials, such as PCPP-SA and PPF. By altering the molecular weight, monomer composition, and crosslinking method, scientists can tailor the crystallinity, hydrophilicity, and degradability of these polymers. In combination with innovative processing methods, this ability allows engineers to design scaffolds and microparticles with specific drug-release kinetics. In order to reduce the minimum effective drug loading within these systems, engineers, biologists, and physicians must now work together to assess how the time course of protein release affects tissue repair. By optimizing the interplay between both the material and the bioactive components of these drug-delivery systems, the field of tissue engineering will undoubtedly revolutionize the treatment of bone degenerative disorders.

Acknowledgements The work on drug delivery for bone tissue engineering was supported by the National Institutes of Health (R01 AR48756). T.A.H. also acknowledges financial support from a Whitaker Foundation Graduate Fellowship.

References

1. Hutmacher DW (2000) Scaffolds in tissue engineering bone and cartilage. Biomaterials 21:2529–2543
2. Beck LS, Deguzman L, Lee WP, Xu Y, McFatridge LA, Gillett NA, Amento EP (1991) Rapid publication. TGF-beta 1 induces bone closure of skull defects. J Bone Miner Res 6:1257–1265

3. Ferguson D, Davis WL, Urist MR, Hurt WC, Allen EP (1987) Bovine bone morphogenetic protein (bBMP) fraction-induced repair of craniotomy defects in the rhesus monkey (Macaca speciosa). Clin Orthop 251–258

4. Mori M, Isobe M, Yamazaki Y, Ishihara K, Nakabayashi N (2000) Restoration of segmental bone defects in rabbit radius by biodegradable capsules containing recombinant human bone morphogenetic protein-2. J Biomed Mater Res 50:191–198

5. Lucas PA, Laurencin C, Syftestad GT, Domb A, Goldberg VM, Caplan AI, Langer R (1990) Ectopic induction of cartilage and bone by water-soluble proteins from bovine bone using a polyanhydride delivery vehicle. J Biomed Mater Res 24:901–911

6. Winn SR, Schmitt JM, Buck D, Hu Y, Grainger D, Hollinger JO (1999) Tissue-engineered bone biomimetic to regenerate calvarial critical-sized defects in athymic rats. J Biomed Mater Res 45:414–421

7. Andriano KP, Tabata Y, Ikada Y, Heller J (1999) In vitro and in vivo comparison of bulk and surface hydrolysis in absorbable polymer scaffolds for tissue engineering. J Biomed Mater Res 48:602–612

8. Elisseeff J, Anseth K, Sims D, McIntosh W, Randolph M, Langer R (1999) Transdermal photopolymerization for minimally invasive implantation. Proc Natl Acad Sci USA 96:3104–3107

9. Babensee JE, McIntire LV, Mikos AG (2000) Growth factor delivery for tissue engineering. Pharm Res 17:497–504

10. Anseth KS, Metters AT, Bryant SJ, Martens PJ, Elisseeff JH, Bowman CN (2002) In situ forming degradable networks and their application in tissue engineering and drug delivery. J Controlled Release 78:199–209

11. Payne RG, McGonigle JS, Yaszemski MJ, Yasko AW, Mikos AG (2002) Development of an injectable, in situ crosslinkable, degradable polymeric carrier for osteogenic cell populations. Part 3. Proliferation and differentiation of encapsulated marrow stromal osteoblasts cultured on crosslinking poly(propylene fumarate). Biomaterials 23:4381–4387

12. Payne RG, McGonigle JS, Yaszemski MJ, Yasko AW, Mikos AG (2002) Development of an injectable, in situ crosslinkable, degradable polymeric carrier for osteogenic cell populations. Part 2. Viability of encapsulated marrow stromal osteoblasts cultured on crosslinking poly(propylene fumarate). Biomaterials 23:4373–4380

13. Payne RG, Yaszemski MJ, Yasko AW, Mikos AG (2002) Development of an injectable, in situ crosslinkable, degradable polymeric carrier for osteogenic cell populations. Part 1. Encapsulation of marrow stromal osteoblasts in surface crosslinked gelatin microparticles. Biomaterials 23:4359–4371

14. Ito Y, Inoue M, Liu SQ, Imanishi Y (1993) Cell growth on immobilized cell growth factor 6. Enhancement of fibroblast cell growth by immobilized insulin and/or fibronectin. J Biomed Mater Res 27:901–907

15. Liu SQ, Ito Y, Imanishi Y (1993) Cell growth on immobilized cell growth factor 9. Covalent immobilization of insulin, transferrin, and collagen to enhance growth of bovine endothelial cells. J Biomed Mater Res 27:909–915

16. Shin H, Jo S, Mikos AG (2002) Modulation of marrow stromal osteoblast adhesion on biomimetic oligo[poly(ethylene glycol) fumarate] hydrogels modified with Arg-Gly-Asp peptides and a poly(ethylene glycol) spacer. J Biomed Mater Res 61:169–179

17. Shin H, Jo S, Mikos AG (2003) Biomimetic materials for tissue engineering. Biomaterials 24:4353–4364

18. Shin H, Zygourakis K, Carson-Farach MC, Yaszemski MJ, Mikos AG (2004) Attachment, proliferation, and migration of marrow stromal osteoblasts cultured on

biomimetic hydrogels modlified with an osteopontin-derived peptide. Biomaterials 25:895–906

19. Rodan GA, Martin TJ (2000) Therapeutic approaches to bone diseases. Science 289:1508–1514

20. Wang J, Chow D, Heiati H, Shen WC (2003) Reversible lipidization for the oral delivery of salmon calcitonin. J Controlled Release 88:369–380

21. Mestiri M, Benoit JP, Hernigou P, Devissaguet JP, Puisieux F (1995) Cisplatin-loaded poly(methyl methacrylate) implants: a sustained drug delivery system. J Controlled Release 33:107–113

22. Downes S (1995) Growth hormone release from biomaterials. In: Wise DL, Trantolo DJ, Altobelli DE, Yaszemski MJ, Gresser JD, Schwartz ER (eds) Encyclopedic handbook of biomaterials and bioengineering, part a: materials. Dekker, New York, pp 1135–1149

23. Seligson D, Henry SL (1993) Newest knowledge of treatment for bone infection: antibiotic-impregnated beads. Clin Orthop 295:2–118

24. Thomson RC, Ishaug SL, Mikos AG, Langer R (1995) Polymers for biological systems. In: Meyers RA (ed) Molecular biology and biotechnology, a comprehensive desk reference. VCH, New York, pp 717–724

25. Burg KJ, Porter S, Kellam JF (2000) Biomaterial developments for bone tissue engineering. Biomaterials 21:2347–2359

26. Gittens SA, Uludag H (2001) Growth factor delivery for bone tissue engineering. J Drug Target 9:407–429

27. Teitelbaum SL (2000) Bone resorption by osteoclasts. Science 289:1504:1508

28. Ducy P, Schinke T, Karsenty G (2000) The osteoblast: a sophisticated fibroblast under central surveillance. Science 289:1501–1504

29. Nimni ME (1997) Polypeptide growth factors: targeted delivery systems. Biomaterials 18:1201–1225

30. Lee SJ (2000) Cytokine delivery and tissue engineering. Yonsei Med J 41:704–719

31. Kim HD, Valentini RF (1997) Human osteoblast response in vitro to platelet-derived growth factor and transforming growth factor-beta delivered from controlled-release polymer rods. Biomaterials 18:1175–1184

32. Wozney JM (2002) Overview of bone morphogenetic proteins. Spine 27:S2–S8

33. Solchaga LA, Cassiede P, Caplan AI (1998) Different response to osteo-inductive agents in bone marrow- and periosteum-derived cell preparations. Acta Orthop Scand 69:426–432

34. Bonewald LF, Mundy GR (1990) Role of transforming growth factor-beta in bone remodeling. Clin Orthop 261:276

35. Border WA, Noble NA (1994) Transforming growth factor beta in tissue fibrosis. N Engl J Med 331:1286–1292

36. Busch O, Solheim E, Bang G, Tornes K (1996) Guided tissue regeneration and local delivery of insulinlike growth factor I by bioerodible polyorthoester membranes in rat calvarial defects. Int J Oral Maxillofac Implants 11:498–505

37. Anusaksathien O, Giannobile WV (2002) Growth factor delivery to re-engineer periodontal tissues. Curr Pharm Biotechnol 3:129–139

38. Thomson RC, Shung AK, Yaszemski MJ, Mikos AG (2000) Polymer scaffold processing. In: Lanza RP, Vacanti LRJ (eds) Principles of tissue engineering. Academic, San Diego, pp 251–262

39. Ratner BD (2002) Reducing capsular thickness and enhancing angiogenesis around implant drug release systems. J Controlled Release 78:211–218

40. Middleton JC, Tipton AJ (2000) Synthetic biodegradable polymers as orthopedic devices. Biomaterials 21:2335–2346
41. Bulpitt P, Aeschlimann D (1999) New strategy for chemical modification of hyaluronic acid: preparation of functionalized derivatives and their use in the formation of novel biocompatible hydrogels. J Biomed Mater Res 47:152–169
42. DeBlois C, Cote MF, Doillon CJ (1994) Heparin-fibroblast growth factor-fibrin complex: in vitro and in vivo applications to collagen-based materials. Biomaterials 15:665–672
43. Yamamoto M, Takahashi J, Tabata Y (2003) Controlled release by biodegradable hydrogels enhances the ectopic bone formation of bone morphogenetic protein. Biomaterials 24:4375–4383
44. Edelman ER, Mathiowitz E, Langer R, Klagsbrun M (1991) Controlled and modulated release of basic fibroblast growth factor. Biomaterials 12:619–626
45. Sakiyama-Elbert SE, Hubbell JA (2000) Development of fibrin derivatives for controlled release of heparin-binding growth factors. J Controlled Release 65:389–402
46. Park YJ, Lee YM, Lee JY, Seol YJ, Chung CP, Lee SJ (2000) Controlled release of platelet-derived growth factor-BB from chondroitin sulfate-chitosan sponge for guided bone regeneration. J Controlled Release 67:385–394
47. Mattioli-Belmonte M, Gigante A, Muzzarelli RA, Politano R, DeBenedittis A, Specchia N, Buffa A, Biagini G, Greco F (1999) N,N-Dicarboxymethyl chitosan as delivery agent for bone morphogenetic protein in the repair of articular cartilage. Med Biol Eng Comp 37:130–134
48. Urist MR, Silverman BF, Buring K, Dubuc FL, Rosenberg JM (1967) The bone induction principle. Clin Orthop 53:243–283
49. Urist MR (1995) Experimental delivery systems for bone morphogenetic protein. In: Wise DL, Trantolo DJ, Altobelli DE, Yaszemski MJ, Gresser JD, Schwartz ER (eds) Encyclopedic handbook of biomaterials and bioengineering, part a: materials. Dekker, New York, pp 1093–1133
50. Gombotz WR, Pankey SC, Bouchard LS, Ranchalis J, Puolakkainen P (1993) Controlled release of TGF-beta 1 from a biodegradable matrix for bone regeneration. J Biomater Sci Polym Ed 5:49–63
51. Appel LE, Balena R, Cortese M, Opas E, Rodan G, Seedor G, Zentner GM (1993) In vitro characterization and in vivo efficacy of a prostaglandin E2/poly(orthoester) implant for bone growth production. J Controlled Release 26:77–85
52. Deschamps AA, Claase MB, Sleijster WJ, de Bruijn JD, Grijpma DW, Feijen J (2002) Design of segmented poly(ether ester) materials and structures for the tissue engineering of bone. J Controlled Release 78:175–186
53. Gopferich A (1996) Mechanisms of polymer degradation and erosion. Biomaterials 17:103–114
54. Kohn J, Langer R (1996) Bioresorbable and bioerodible materials. In: Ratner BD, Hoffman AS, Schoen FJ, Lemons JE (eds) Biomaterials science: an introduction to materials in medicine. Academic, New York, pp 64–72
55. Barrows TH (1986) Degradable implant materials: a review of synthetic absorbable polymers and their applications. Clin Mater 1:233–257
56. Domb AJ, Amselem S, Langer R, Maniar M (1994) Polyanhydride Carriers of Drugs. In: Shalably SW (ed) Biomedical polymers, designed-to-degrade systems. Hanser, New York, pp 69–96
57. Schmidt C, Wenz R, Nies B, Moll F (1995) Antiobiotic in vivo/in vitro release, histocompatibility and biodegradation of gentamicin implants based on lactic acid polymers and copolymers. J Controlled Release 37:83–94

58. Soriano I, Evora C (2000) Formulation of calcium phosphates/poly (D,L-lactide) blends containing gentamicin for bone implantation. J Controlled Release 68:121–134

59. Ramchandani M, Robinson D (1998) In vitro and in vivo release of ciprofloxacin from PLGA 50:50 implants. J Controlled Release 54:167–175

60. Baro M, Sanchez E, Delgado A, Perera A, Evora C (2002) In vitro–in vivo characterization of gentamicin bone implants. J Controlled Release 83:353–364

61. Smith JL, Jin L, Parsons T, Turek T, Ron E, Philibrook CM, Kenley RA, Marden L, Hollinger J, Bostrom MPG, Tomin E, Lane JM (1995) Osseous regeneration in preclinical models using bioabsorbable delivery technology for recominbant human bone morphogenetic protein 2 (rhBMP-2). J Controlled Release 36:183–195

62. Zellin G, Linde A (1997) Importance of delivery systems for growth-stimulatory factors in combination with osteopromotive membranes. An experimental study using rhBMP-2 in rat mandibular defects. J Biomed Mater Res 35:181–190

63. Boyan BD, Lohmann CH, Somers A, Niederauer GG, Wozney JM, Dean DD, Carnes DL Jr, Schwartz Z (1999) Potential of porous poly-D,L-lactide-co-glycolide particles as a carrier for recombinant human bone morphogenetic protein-2 during osteoinduction in vivo. J Biomed Mater Res 46:51–59

64. Isobe M, Yamazaki Y, Mori M, Amagasa T (1999) Bone regeneration produced in rat femur defects by polymer capsules containing recombinant human bone morphogenetic protein-2. J Oral Maxillofac Surg 57:695–698; discussion 699

65. Mayer M, Hollinger J, Ron E, Wozney J (1996) Maxillary alveolar cleft repair in dogs using recombinant human bone morphogenetic protein-2 and a polymer carrier. Plast Reconstr Surg 98:247–259

66. Kirker-Head CA, Gerhart TN, Armstrong R, Schelling SH, Carmel LA (1998) Healing bone using recombinant human bone morphogenetic protein 2 and copolymer. Clin Orthop 205–217

67. Miyamoto S, Takaoka K, Okada T, Yoshikawa H, Hashimoto J, Suzuki S, Ono K (1992) Evaluation of polylactic acid homopolymers as carriers for bone morphogenetic protein. Clin Orthop 274–285

68. Park YJ, Ku Y, Chung CP, Lee SJ (1998) Controlled release of platelet-derived growth factor from porous poly(L-lactide) membranes for guided tissue regeneration. J Controlled Release 51:201–211

69. Lee JY, Nam SH, Im SY, Park YJ, Lee YM, Seol YJ, Chung CP, Lee SJ (2002) Enhanced bone formation by controlled growth factor delivery from chitosan-based biomaterials. J Controlled Release 78:187–197

70. Peter SJ, Miller MJ, Yasko AW, Yaszemski MJ, Mikos AG (1998) Polymer concepts in tissue engineering. J Biomed Mater Res 43:422–427

71. Ishaug SL, Payne RG, Yaszemski MJ, Aufdemorte TB, Bizios R, Mikos AG (1996) Osteoblast migration on poly(alpha-hydroxy esters). Biotechnol Bioeng 50:443–451

72. Ishaug-Riley SL, Crane GM, Gurlek A, Miller MJ, Yasko AW, Yaszemski MJ, Mikos AG (1997) Ectopic bone formation by marrow stromal osteoblast transplantation using poly(DL-lactic-co-glycolic acid) foams implanted into the rat mesentery. J Biomed Mater Res 36:1–8

73. Whang K, Tsai DC, Nam EK, Aitken M, Sprague SM, Patel PK, Healy KE (1998) Ectopic bone formation via rhBMP-2 delivery from porous bioabsorbable polymer scaffolds. J Biomed Mater Res 42:491–499

74. Whang K, Goldstick TK, Healy KE (2000) A biodegradable polymer scaffold for delivery of osteotropic factors. Biomaterials 21:2545–2551

75. Wu XS (1995) Preparation, characterization, and drug delivery applications of microspheres based on biodegradable lactic/glycolic acid polymers. In: Wise DL, Trantolo DJ, Altobelli DE, Yaszemski MJ, Gresser JD, Schwartz ER (eds) Encyclopedic handbook of biomaterials and bioengineering, part a: materials. Dekker, New York, pp 1151–1199

76. Peter SJ, Lu L, Kim DJ, Stamatas GN, Miller MJ, Yaszemski MJ, Mikos AG (2000) Effects of transforming growth factor beta1 released from biodegradable polymer microparticles on marrow stromal osteoblasts cultured on poly(propylene fumarate) substrates. J Biomed Mater Res 50:452–462

77. Oldham JB, Lu L, Zhu X, Porter BD, Hefferan TE, Larson DR, Currier BL, Mikos AG, Yaszemski MJ (2000) Biological activity of rhBMP-2 released from PLGA microspheres. J Biomech Eng 122:289–292

78. Meinel L, Illi OE, Zapf J, Malfanti M, Peter Merkle P, Gander B (2001) Stabilizing insulin-like growth factor-I in poly(D,L-lactide-co-glycolide) microspheres. J Controlled Release 70:193–202

79. Lu L, Stamatas GN, Mikos AG (2000) Controlled release of transforming growth factor beta1 from biodegradable polymer microparticles. J Biomed Mater Res 50:440–451

80. Richardson TP, Peters MC, Ennett AB, Mooney DJ (2001) Polymeric system for dual growth factor delivery. Nat Biotechnol 19:1029–1034

81. Hedberg EL, Tang A, Crowther RS, Carney DH, Mikos AG (2002) Controlled release of an osteogenic peptide from injectable biodegradable polymeric composites. J Controlled Release 84:137–150

82. Marra KG, Szem JW, Kumta PN, DiMilla PA, Weiss LE (1999) In vitro analysis of biodegradable polymer blend/hydroxyapatite composites for bone tissue engineering. J Biomed Mater Res 47:324–335

83. Ekholm M, Hietanen J, Lindqvist C, Rautavuori J, Santavirta S, Suuronen R (1999) Histological study of tissue reactions to epsilon-caprolactone-lactide copolymer in paste form. Biomaterials 20:1257–1262

84. Kellomaki M, Niiranen H, Puumanen K, Ashammakhi N, Waris T, Tormala P (2000) Bioabsorbable scaffolds for guided bone regeneration and generation. Biomaterials 21:2495–2505

85. Jazayeri MA, Nichter LS, Zhou ZY, Wellisz T, Cheung DT (1994) Comparison of various delivery systems for demineralized bone matrix in a rat cranial defect model. J Craniofacial Surg 5:172–178

86. Bennett S, Connolly K, Lee DR, Jiang Y, Buck D, Hollinger JO, Gruskin EA (1996) Initial biocompatibility studies of a novel degradable polymeric bone substitute that hardens in situ. Bone 19:101S–107S

87. Miyamoto S, Takaoka K, Okada T, Yoshikawa H, Hashimoto J, Suzuki S, Ono K (1993) Polylactic acid-polyethylene glycol block copolymer. A new biodegradable synthetic carrier for bone morphogenetic protein. Clin Orthop 333–343

88. Saito N, Okada T, Horiuchi H, Murakami N, Takahashi J, Nawata M, Ota H, Miyamoto S, Nozaki K, Takaoka K (2001) Biodegradable poly-D,L-lactic acid-polyethylene glycol block copolymers as a BMP delivery system for inducing bone. J Bone Joint Surg Am 83-A Suppl 1:S92–98

89. Saito N, Okada T, Horiuchi H, Ota H, Takahashi J, Murakami N, Nawata M, Kojima S, Nozaki K, Takaoka K (2003) Local bone formation by injection of recombinant human bone morphogenetic protein-2 contained in polymer carriers. Bone 32:381–386

90. Saito N, Okada T, Horiuchi H, Murakami N, Takahashi J, Nawata M, Ota H, Nozaki K, Takaoka K (2001) A biodegradable polymer as a cytokine delivery system for inducing bone formation. Nat Biotechnol 19:332–335

91. Tessmar J, Mikos AG, Gopferich A (2003) The use of poly(ethylene glycol)-block-poly(lactic acid) derived copolymers for the rapid creation of biomimetic surfaces. Biomaterials 24:4475–4486

92. Gao TJ, Kousinioris NA, Wozney JM, Winn S, Uludag H (2002)Synthetic thermoreversible polymers are compatible with osteoinductive activity of recombinant human bone morphogenetic protein 2. Tissue Eng 8:429–440

93. Uludag H, Norrie B, Kousinioris N, Gao T (2001) Engineering temperature-sensitive poly(N-isopropylacrylamide) polymers as carriers of therapeutic proteins. Biotechnol Bioeng 73:510–521

94. Hacker M, Tessmar J, Neubauer M, Blaimer A, Blunk T, Gopferich A, Schulz MB (2003) Towards biomimetic scaffolds: Anhydrous scaffold fabrication from biodegradable amine-reactive diblock copolymers. Biomaterials 24:4459–4473

95. West JL, Hubbell JA (1995) Photopolymerizable hydrogel materials for drug delivery applications. React Polym 25:139–147

96. Mason MN, Metters AT, Bowman CN, Anseth K (2001) Predicting controlled-release behavior of degradable PLA-b-PEG-b-PLA hydrogels. Macromol 34

97. Burdick JA, Mason MN, Hinman AD, Thorne K, Anseth KS (2002) Delivery of osteoinductive growth factors from degradable PEG hydrogels influences osteoblast differentiation and mineralization. J Controlled Release 83:53–63

98. Jo S, Shin H, Shung AK, Fisher JP, Mikos AG (2001) Synthesis and characterization of oligo(poly(ethylene glycol) fumarate) macromer. Macromolecules 34:2839–2844

99. Shin H, Ruhe PQ, Mikos AG, Jansen JA (2003) In vivo bone and soft tissue response to injectable biodegradable oligo(poly(ethylene glycol) fumarate) hydrogels. Biomaterials 24:3201–3211

100. Temenoff JS, Athanasiou KA, LeBaron RG, Mikos AG (2002) Effect of poly(ethylene glycol) molecular weight on tensile and swelling properties of oligo(poly(ethylene glycol) fumarate) hydrogels for cartilage tissue engineering. J Biomed Mater Res 59:429–437

101. Holland TA, Tabata Y, Mikos AG (2003) In vitro release of transforming growth factor-$\beta 1$ from gelatin microparticles encapsulated in biodegradable, injectable oligo(poly(ethylene glycol) fumarate) hydrogels. J Controlled Release 91:299–313

102. Mathiowitz E, Langer R (1987) Polyanhydride microspheres as drug carriers I Hot-melt microencapsulation. J Controlled Release 5:13–22

103. Ron E, Turek T, Mathiowitz E, Chasin M, Hageman M, Langer R (1993) Controlled release of polypeptides from polyanhydrides. Proc Natl Acad Sci USA 90:4176–4180

104. Nelson CL, Hickmon SG, Skinner RA (1992) The treatment of experimental osteomyelitis by a local biodegradable antibiotic delivery system. Orthopaedic Research Society 38th Annual Meeting, p 431

105. Mathiowitz E, Kline D, Langer R (1990) Morphology of polyanhydride microsphere delivery systems. Scanning Microsc 4:329–340

106. Anseth KS, Shastri VR, Langer R (1999) Photopolymerizable degradable polyanhydrides with osteocompatibility. Nat Biotechnol 17:156–159

107. Timmer MD, Ambrose CG, Mikos AG (2003) Evaluation of thermal- and photo-crosslinked biodegradable poly(propylene fumarate)-based networks. J Biomed Mater Res 66A:811–818

108. Fisher JP, Timmer MD, Holland TA, Dean D, Engel PS, Mikos AG (2003) Photoinitiated cross-linking of the biodegradable polyester poly(propylene fumarate) Part I: determination of network structure. Biomacromolecules 4:1327–1334

109. Fisher JP, Holland TA, Dean D, Engel PS, Mikos AG (2001) Synthesis and properties of photocross-linked poly(propylene fumarate) scaffolds. J Biomater Sci Polym Ed 12:673–687

110. Fisher JP, Vehof JW, Dean D, van der Waerden JP, Holland TA, Mikos AG, Jansen JA (2002) Soft and hard tissue response to photocrosslinked poly(propylene fumarate) scaffolds in a rabbit model. J Biomed Mater Res 59:547–556

111. Vehof JW, Fisher JP, Dean D, van der Waerden JP, Spauwen PH, Mikos AG, Jansen JA (2002) Bone formation in transforming growth factor beta-1-coated porous poly(propylene fumarate) scaffolds. J Biomed Mater Res 60:241–251

112. Hedberg E, Kroese-Deutman H, Lemoine J, Shih C, Crowther R, Carney D, Liebschner M, Mikos A, Jansen J (2003) In vivo osteogenesis in response to the controlled release of TP508 from biodegradable polymeric scaffolds. Controlled Release Society 30th Annual Meeting Proceedings, p 90

113. Uludag H, Fan XD (2000) Controlled drug delivery: designing technologies for the future. In: Park K, Mrsny R (eds) Drug delivery in the 21st century. American Chemical Sociey, Washington, DC, pp 253–262

114. Hahn M, Gornitz E, Dautzenberg H (1998) Synthesis and properties of ionically modified polymers with LCST behavior. Macromol 31:5616–5623

115. Cordoba F, Dong SS, Robinson M, Strates BS, Nimni ME (1993) Effect of microcrystalline hydroxyapatite on bone marrow stromal cell osteogenesis. Orthopaedic Research Society 39th Annual Meeting, p 102

116. McGill JJ, Strates BS, McGuire MH (1991) Stimulation of osteogenesis by PDGF and TGF-beta adsorbed on microcrystalline hydroxyapatite. J Bone Miner Res 6:503

117. Strates BS, Kilaghbian V, Nimni ME, McGuire MH, Petty RW (1992) Enhanced periosteal osteogenesis induced by rTGF-beta1 absorbed on microcrystals of hydroxyapatite. Orthopaedic Research Society 38th Annual Meeting, p 591

118. Holland TA, Tessmar J, Tabata Y, Mikos AG (2004) Transforming growth factor-beta1 release from oligo(poly(ethylene gylcol) fumarate) hydrogels in conditions that model the cartilage wound healing environment. J Controlled Release 94:101–114

119. Zhang R, Ma PX (1999) Poly(alpha-hydroxyl acids)/hydroxyapatite porous composites for bone-tissue engineering I Preparation and morphology. J Biomed Mater Res 44:446–455

120. Peter SJ, Kim P, Yasko AW, Yaszemski MJ, Mikos AG (1999) Crosslinking characteristics of an injectable poly(propylene fumarate)/beta-tricalcium phosphate paste and mechanical properties of the crosslinked composite for use as a biodegradable bone cement. J Biomed Mater Res 44:314–321

121. Vogelson C, Koide Y, Barron A (2000) Fiber reinforced epoxy resin composite materials using carboxylate-alumoxanes as cross-linking agents. Mater Res Soc Symp Proc 581:369–374

122. Vogelson C, Koide Y, Alemany L, Barron A (2000) Inorganic-organic hybrid and composite resin materials using carboxylate-alumoxanes as functionalized crosslinking agents. Chem Mater 12:795–904

123. Illi OE, Feldmann CP (1998) Stimulation of fracture healing by local application of humoral factors integrated in biodegradable implants. Eur J Pediatr Surg 8:251–255

124. Barratt G, Couarraze G, Couvreur P, Dubernet C, Fattal E, Gref R, Labarre D, Legrand P, Ponchel G, Vauthier C (2002) Polymeric micro- and nanoparticles as drug carriers. In: Dumitriu S (ed) Polymeric biomaterials. Dekker, New York, pp 753–781

125. Gibaud S, Rousseau C, Weingarten C, Favier R, Douay L, Andreux JP, Couvreur P (1998) Polyalkylcyanoacrylate nanoparticles as carriers for granulocyte-colony stimulating factor (G-CSF). J Controlled Release 52:131–139
126. Cooke MN, Fisher JP, Dean D, Rimnac C, Mikos AG (2003) Use of stereolithography to manufacture critical-sized 3D biodegradable scaffolds for bone ingrowth. J Biomed Mater Res 64B:65–69

Adv Biochem Engin/Biotechnol (2006) 102: 187–238
DOI 10.1007/10_013
© Springer-Verlag Berlin Heidelberg 2006
Published online: 15 July 2006

Biopolymer-Based Biomaterials as Scaffolds for Tissue Engineering

James Velema · David Kaplan (✉)

Department of Biomedical Engineering, Tufts University, 4 Colby Street,
Medford, MA 02155, USA
david.kaplan@tufts.edu

Abstract Biopolymers as biomaterials and matrices in tissue engineering offer important options in control of structure, morphology and chemistry as reasonable substitutes or mimics of extracellular matrix systems. These features also provide for control of material functions such as mechanical properties in gel, fiber and porous scaffold formats. The inherent biodegradability of biopolymers is important to help regulate the rate and extent of cell and tissue remodeling in vitro or in vivo. The ability to genetically redesign these polymer systems to bioengineer appropriate features to regulate cell responses and interactions is another important feature that offers both fundamental insight into chemistry–structure–function relationships as well as direct utility as biomaterials. Biopolymer matrices for biomaterials and tissue engineering can directly influence the functional attributes of tissues formed on these materials and suggest they will continue play an increasingly important role in the field.

Keywords Biopolymers · Biodegradable · Protein · Polysaccharide · Polyester · Biosynthesis

1
Introduction

Reconstructive medicine is undergoing a major transition with regard to the materials used as replacements for diseased or damaged tissue. This transition is driven by a number of factors, including the observation that synthetic implants ultimately fail the long-term test of biological compatibility, the population as a whole is living longer and thus implants need to better integrate into the surrounding tissue to avoid recurring replacement, and there is a continued demand for more rapid restoration of full function. These issues have fostered a change in approach [1, 2]. Instead of continuing the search for inert materials, the focus in the field of tissue engineering has been on the design and engineering of biodegradable materials that interact with and facilitate the rebuilding of native tissue during their residence time in the body. A common approach utilizes the biomaterial as a temporary scaffold for the delivery and integration of cells and/or growth factors at the repair site [3, 4]. This role requires that the scaffold material provide sufficient structure and function (e.g., mechanical support) during the remodeling process. Degradability must be predictable so that tissue regeneration occurs at a rate that matches the degradation of the scaffold.

In addition to a primary function to provide structural support, it is desirable for the biomaterial to promote and maintain an environment that enables appropriate cellular adhesion, growth and differentiation, leading to targeted phenotypic expression that will ultimately lead to new tissue formation and function. Synthetic biodegradable polymers have been used extensively as supports for cell growth, but only recently have attempts been made to supplement these materials with bioactive molecules to stimulate or modulate the remodeling process (for recent reviews, see [3, 5]). A separate approach to the needs in the field of tissue engineering involves the use of naturally derived polymers with their intrinsic ability to stimulate tissue growth. Many of these molecules are components of the extracellular matrix, thus they are polymers that ordinarily interact with cells. Since the extracellular matrix plays an important role in regulating aspects of cell division, adhesion, cell motility, differentiation and migration, natural biopolymers serve as important starting points in attempts to recapitulate natural properties of tissues in a controlled fashion either in vitro or in vivo.

A gross comparison of synthetic and natural polymers in terms of versatility and control of structure and function leads to the conclusion that the design space available with biopolymers exceeds synthetic polymers by a large margin, using whatever measure one can impose. Thus, biopolymers (e.g., proteins, polysaccharides, nucleic acids) offer diverse chemistries (e.g., 20 different side chains in the case of amino acids), highly selective coupling chemistries (regioselective and stereoselective) and chain-length control in the case of those biopolymers generated from template-based synthesis (e.g., proteins, nucleic acids). Aside from diversity and control of chemistry, the attributes mentioned above in terms of control of links between monomers leads to a unique and important suite of features for biopolymers, predictable control of intra- and inter-chain interactions, leading to well-defined structural hierarchies that propagate these chain–chain interactions up length scales through the mesoscopic to the macroscopic (scaffold, tissue) level. This process, driven by many weak forces (e.g., hydrophobic, hydrogen bonding) drives hierarchical assembly in biopolymers and leads to the unique and important diversity in structures and mechanical functions in tissue ECM. These features of chemistry and structural assembly/hierarchy are important aspects attributable to the use of biopolymers as scaffolds.

Three classes of natural polymers, polysaccharides, proteins, and polyhydroxyalkanoates (PHAs), have been investigated as biomaterials (Table 1). Much of the interest in polysaccharides as biomaterials stems from their structural and biochemical similarity to glycosaminoglycans (GAG), such as hyaluronate, dermatan sulfate, chondroitin sulfate, keratin sulfate, of natural vertebrate ECM. Hyaluronate is the longest chain of all GAG molecules (500 to several thousand unbranched sugars) and has been the most intensively studied as a tissue-engineering scaffold due to its relative ease of preparation and modification. Nevertheless, preliminary investigations have been made into other GAGs, particularly chondroitin sulfate, which may be used as components of scaffolds to promote vascularization and rapid regeneration of soft tissues [13]. The high charge density and hydrophilicity of polysaccharides facilitates the formation of highly hydrated (> 30% water content) gels or *hydrogels*, which are ideal for cell encapsulation [23]. The highly hydrophilic surface character of these materials disfavors adsorption by extracellular proteins and receptor-mediated cell adhesion in favor of cell–cell interactions.

Table 1 Summary of the applications of naturally derived biomaterials (adapted from [6])

Natural polymer	Current Uses as biomaterials	Use in tissue engineering	Tissue engineering Refs.
A) Polysaccharides and derivatives			
Alginate	Immobilization matrix for cells and enzymes; controlled release of bioactive substances; injectable microspheres	Reviewed herein	Reviewed herein
Agarose	Immobilization matrix	Load-sensitive construct for cartilage development in vitro	Mauck et al. 2003 [7] Mauck et al. 2000 [8]
Cellulose	Cell immobilization, drug delivery, dialysis membranes	Injection-molded porous scaffolds for regeneration of load-bearing tissues	Gomes et al. 2001 [9]
Carboxy-methyl-cellulose	Oolyelectrolyte complexes for immunoisolation	Injectable scaffold for repair of brain defects	Tate et al. 2001 [10]
Chitosan	Controlled delivery systems (ex. gels, membranes, microspheres)	Reviewed herein	Reviewed herein
Dextran	Plasma expander, drug carrier	Amine crosslinked sponges for cell delivery	Ehrenfreund-Kleinman et al. 2002 [11]
Heparin	Anti-thrombotic, anti-coagulant	Vinylated heparin for photocurable scaffolds	Matsuda et al. 2002 [12]
Hyaluronic acid	Lubricant; visco-supplement for knee and eye surgery; post-surgical adhesion prevention	Reviewed herein	Reviewed herein
Other GAGs: heparan sulfate; chondroitin sulfate; dermatan sulfate	Components of collagen-GAG burn dressings for wound healing; skin replacements	Attachment to collagen matrices improves scaffold vascularization	Pieper et al. 2002 [13]; Pieper, van Wachem et al. 2000 [14]; Pieper, Hafmans et al. 2000 [15]
B) Proteins and protein-based polymers			
Collagen	Absorbable sutures, sponge wound and burn dressing, drug-delivery microspheres, soft-tissue augmentation	Reviewed herein	Reviewed herein

Table 1 (continued)

Natural polymer	Current uses as biomaterials	Use in tissue engineering	Tissue engineering Refs.
Elastin	None	Reviewed herein	Reviewed herein
Fibrin/ fibrinogen	Glues or sealants as soft-tissue adhesives or wound dressings	Gels for cartilage or vascular reconstruction	Passaretti 2001 [16] Ye et al. 2000 [17]
Fibronectin	Coatings on synthetic materials to improve cell adhesion	Fibronectin–fibrinogen cables for wound repair	Underwood 2001 [18]
Silk	Absorbable sutures	Reviewed herein	Reviewed herein
C) Polyhydroxyalkanoates			
Poly-hydroxy-butyrate (PHB)	Controlled drug release, sutures, artificial skin	Nerve regeneration conduits; heart valves; vessel augmentation; vascular grafts	Novikov 2002 [19]; Sodian et al. 2000 [20]; Stock, Sakamoto et al. 2000 [21] Stock, Nagashima, et al. 2000 [22]

Along with their low crystallinity, these properties impart low cytotoxicity and minimal inflammatory response. In this review, alginate, chitosan, and hyaluronate will be discussed, polysaccharides that have been most intensively investigated as potential biomaterial scaffolds.

Proteins are another family of natural polymers that have attracted interest as tissue scaffold materials. Fibrous proteins have been of particular interest due to the highly repetitive secondary and tertiary peptide structures that confer mechanical rigidity to these molecules. Other naturally derived proteins have been investigated to a lessor extent (Table 1). Fibrin for example, has been exploited for its ECM-protein synthesis, promoting properties in the form of fibrin gels. However, long-term implantation is hampered by the intrinsic instability of the molecule. Several chemical inhibitors of fibrinolysis have achieved limited success in stabilizing fibrin cell scaffolds [16, 17, 24, 25].

A third group of natural polymers, the polyhydroxyalkanoates (PHAs), have also recently received attention for their potential in tissue engineering. These biological polyesters, comprised of component hydroxyl fatty acid monomers, are normally produced as dense granules in many bacteria. Their natural function is thought to be intracellular storage of excess nutrients in a manner analogous to starch and glycogen [26]. The accumulation of these distinct granules during fermentation also allows for ease of purification. Transgenic E. coli containing the operon of six enzymes for P4HB synthesis can produce yields of polymer in excess of 50 grams per liter and polymer of high molecular weight [27].

Once extracted from their bacterial source, PHAs have thermoplastic properties that facilitate extrusion, injection molding, solution casting and other melt-processing techniques similar to those developed by the plastics industry. PHAs with short or no fatty acyl side-chain substituents have been noted for good strength and elasticity coupled (up to 1000% extensibility and 50 MPa tensile strength in the case a homopolymer of 4-hydroxybutyrate (P4HB)) [27]. These materials have excellent potential as scaffolds for heart valves, vascular grafts, stents, and other vascular applications where resilience to cyclic stress and high burst strength are required for durability [20–22]. Furthermore, their hydrophobicity contributes to gradual degradation over several months to the component hydroxyacid monomers in vivo by surface hydrolysis. These degradation products are weakly acidic metabolites found in all tissues, and the small amounts released are rapidly cleared from the body. At present, the long residence time of the polymers limits their use to applications requiring slow remodeling of the scaffold, but copolymerization with one of many available hydroxyacids may further extend the biodegradation time and range of mechanical properties [28].

This review will summarize recent studies with select biological polysaccharides (alginate, chitosan, and hyaluronan) and fibrous proteins (collagen, silk fibroin, and elastin). The polyhydroxyalkanoates (PHAs), being more chemically similar to synthetic polyesters [polyglycolic acid (PGA) and poly-lactic acids (PLA)], will not be discussed in further detail here (for recent reviews see [27, 28]). The variety of fabrication techniques required to produce porous tissue scaffolds from these materials will be discussed. Emphasis will be placed on how these molecules meet criteria considered important for successful implementation as tissue-engineered scaffolds and biomaterials. Current attempts at tailoring these materials to the engineering of specific tissue types will also be discussed.

2
Chemical and Structural Properties

2.1
Biological Polysaccharides

2.1.1
Alginate

Alginate is a salt form of alginic acid, which is the primary structural polymer of brown algae or seaweed. The physicochemical properties of this biomaterial are primarily derived from the copolymer composition of β-(1-4)-linked disaccharide repeating unit composed of the uronate monosaccharides D-Mannuronic acid and α-L-Glucuronic acid. The relative amount and distribution of these negatively-charged monomers in alternating or repeating blocks of M units or G units depends on the species and age of the seaweed from which the polymer is derived [29]. The ability of the material to form hydrogels is dependent on the presence of multivalent cations which crosslink individual alginate strands through ionic bonds with the carboxylate groups of G units in the adjacent polymers. The highly hydrated (up to 99%) nature of these anionic compositions makes the surface character of the material highly hydrophilic.

2.1.2
Chitosan

Chitosan is a positively charged polysaccharide composed of $\beta(1\text{-}4)$-linked D-glucosamine (GlcN) monosaccharides with randomly interspersed N-acetylglucosamine (GlcNac). Chitosan is a derivative of chitin, the primary structural component of arthropod exoskeletons, wherein the majority of GlcNac sugars are replaced with glucosamine (Glc). The gelation of chitosan is due to the formation of inter-strand hydrogen bonds between free amines groups in adjacent Glc sugars [30]. This is in turn dependent on pH, as protonation of the amine groups reduces the extent of hydrogen bonding. The polymer normally gels under alkaline conditions (> pH 7) but is completely soluble at pH < 5. Commercially available chitosan preparations can be quite variable in composition, with degrees of deacetylation ranging from 50 to 90% and molecular weights of 300–1000 kD.

2.1.3
Hyaluronate

While chitosan and alginate are absent in mammalian tissues, hyaluronate (also known as hyaluronic acid or hyaluronan) is an essential glycoasminoglycan (GAG) of the ECM in all mammals. The chemical composition consists of alternately repeating 1,4-linked disaccharide units of negatively charged 1,3-linked monosaccharides glucuronic acid (Glu) and N-acetyl glucosamine (GlcNac) [31]. In the vertebrate body, unbranched chains of 500 to several thousand disaccharides units are attached to core protein structures as massive proteoglycan aggregates. Hyaluronic acid (HA) preparations from tissues are polydisperse with respect to molecular mass. Usually the extracted HA has an average molecular weight of several million Daltons. The solubility and high water-binding capacity of the HA polymer forms a highly viscous solution that imbues synovial fluid and cartilage with their important lubricant and viscoelastic properties. Other functions include regulation of water balance, selective sieving of plasma and matrix protein diffusion, and an ECM depot for active molecules such as growth factors [32].

The rapid resorption time and short residence time of purified HA limits its use and several attempts have been made to modify its molecular structure to obtain more-stable solid materials. Crosslinked materials by covalently coupled HA chains such as Hylans, can be generated with chemical coupling reagents. The chemistries include biscarbodiimides, carbodiimides, polyfunctional epoxides or HA derivatives that are first functionalized with hydrazines, aldehydes, or amines by modification of specific functional groups on the polysaccharide (eg. carboxyl, hydroxyl, N-acetyl groups) using chemical reactions such as esterification, sulphatation, amidation, etc. [33–35]. HA modified with glycidyl methacrylate can also be photopolymerized to obtain hydrogel scaffolds with a range of swelling and degradation rates [36]. Of these large groups of modified HA molecules, the HA esters have been the most extensively investigated for potential use in tissue engineering [31]. In these derivatives, the free carboxyl group of glucuronic acid is esterified with an aliphatic or aryl alcohol. A broad variety of polymers can be subsequently generated either by changing the type of ester group introduced of extent of esterification.

2.2
Fibrous Proteins

2.2.1
Collagen

Collagen is the most widely utilized natural polymer for biomedical applications. The utility of collagen as a structural biomaterial is reflected by the fact that it is the most common protein in vertebrates. Its rigid right-handed triple-helical rod structure, also known as tropocollagen, is comprised of three left-handed alpha helices of approximately 1000 amino acids in length. The collagen alpha helices contain extensive repeats of the tripeptide sequence Gly – X – Y, in which proline, lysine, and the post-translationally modified amino acids hydroxyproline and hydroxylysine are most commonly found in the X and Y residue positions. Glycine is necessarily present at every third position in the molecule to permit close packing of the alpha helices [37]. Extensive hydrogen bonding between hydroxyprolines and glycines on adjacent polypeptide backbones confers stability to the structure. Of the 19 types of collagen found in the human body, Types I, II and III are the most abundant and most fibrous [38]. Self-aggregation and covalent crosslinking of one or more of these three collagen types facilitates higher-order packing into hexagonal parallel arrays or microfibrils which can be further assembled into sheets, bundles, and other configurations to form skin, bone and tendon (mainly Type I), cartilage (mainly Type II), and blood vessels (mainly Type III) [39]. In native collagen crosslinking of lysine and hydroxylysine residues is accomplished by the enzyme lysyl oxidase, while in biomaterial applications physical methods such as gamma irradiation or chemical crosslinking agents such as glutaraldehyde or carbodiimide must be employed to artificially crosslink the material to reduce rates of biodegradation, increase tensile strength and decrease solubility [40].

2.2.2
Silk

The family of silks is heterogeneous in chemical and physical composition, reflecting evolutionary adaptation to a variety of functions [41]. Of all silk fibers, those from the cocoon of the silkworm (*Bombyx mori*) and the dragline of the golden orb weaver spider (*Nephila clavipes*) have been the most extensively studied. Most fibers of silkworm or spiders consist of component fibroin or spidroin proteins that vary in molecular weight between species and silk types. In the case of spider dragline silk, the secondary structure of these proteins when assembled consists of a block copolymer-like arrangement of large segments with up to 10 repeats of glycine-rich repeat sequences interspersed with small crystalline regions of poly-alanine (alanine-glycine repeats in the case of *B. mori* fibroin) sequences [42, 43]. Polyalanine domains form antiparallel β-pleated sheet regions that are thought to correspond to crystalline domains that noncovalently crosslink the individual fibroin molecules and provide silks with the tensile strength [44, 45]. The small, nonpolar alanine side chains extend perpendicularly from the protein backbone and allow the sheets to pack closely together, stabilized by noncovalent van der Waals and hydrophobic interactions. The glycine-rich sequences, by contrast, are loosely conserved motifs of which Ala – Gly – X (where X is serine or tyrosine) is the most common in *B. mori*, and Gly – Gly – X, Gly – Pro – Gly – Gly – X, or Gly – Pro – Gly – Gln – Gln, (where X is often tyrosine glutamine, or lysine) are found in *N. clavipes* [46, 47]. These sequences form noncrystalline domains that are primarily responsible for the more elastomeric properties of spider silks. Recent solid-state NMR studies suggested that these domains are not amorphous but adopt a preferred and highly ordered 3_{10}-helical [44] or β-turn spiral [46] domains. The silk drawing speed influences the orientation and packing of the crystalline domains that are strengthened by hydrophobic interactions and hydrogen bonding. Polymer chains are organized predominantly parallel to axis of the silk fiber, resulting in a fiber that is strong and relatively inextensible.

The formation of higher-order fibril structures from a highly concentrated aqueous dope solution of silk fibroin protein involves the formation of micelle-like structures as water is gradually removed [48]. The partitioning of hydrophilic and hydrophobic domains of the protein promotes solubility of the micelles in water while preventing premature β-sheet formation. Spiders and silkworms carefully control the removal of water until the concentration of protein increases to coalesce the soft micelles into gel states and liquid-crystalline structures. The application of sheer stress during fiber spinning induces alignment and β-sheet formation.

2.2.3
Elastin

As with collagen, elastic fibers are an important component of mammalian connective tissue, especially those tissues that must expand and contract repetitively without failure, such as the aorta and lung. This property is derived primarily due to the 75-kDa elastomeric protein elastin, which can withstand significant deformation without rupture while reverting to its original length when stress is removed. Elastin consists of a highly repetitive hydrophobic sequence, including valine and glycine (GVGVP and variations) that is repeated throughout the protein chain in regions described as *elastomeric domains* [43]. Like collagen, glycine accounts for one third of all amino acids in the protein, but its distribution is more random. Spectroscopic studies by Urry and others have indicated that type II β-turns are a prominent feature in these elastomeric domains and that the repetition of this sequence results in a loosely helical *β-spiral* structure which can support large-amplitude low-frequency vibrations [49, 50]. Other studies, imply a more isolated arrangement of β-turns which are labile and interconvertible and provide for the intrinsic entropy of elastin [51]. Whatever model accounts for its properties, it is established that a limited number of intramolecular hydrogen bonds occur between these structures, and the limited constraint to rotational motion allows elastin molecules to slide extensively over each other. Despite its hydrophobicity, the polymer is extensively hydrated when in the unstretched relaxed state and water acts as a plasticizer to increase its entropy further [52].

Like silk, the elastomeric properties of elastin are primarily due to the alignment and ordering of the highly repeated sequences upon stretching. This stretching decreases the hydration. Elastin is more resilient than silk, with a recoil force that is higher (90% of input energy), than for silk (35%). Interspersed between the elastomeric domains are small crystalline regions of primarily α-helical and Ala-rich regions containing Lys residues that are crosslinked by lysyl oxidase. These *crosslinking domains* link the lysine chains of four tropoelastin molecules into an insoluble elastin polymer through covalent structures called desmosine junctions [53]. Crosslinks occur every 65 – 70 amino acids, forming a random, highly entropic network of elastin molecules, and providing the resistance to deformation upon loading. Stress orders the network and decreases its entropy by limiting the conformational freedom of the chains and providing the restoring force [43].

Many of the useful properties of natural elastin have been captured in genetically engineered elastin-like polypeptides (ELPs) containing tetra-, penta-, and hexapeptide repeat motifs [54–56]. Only variations of the pentapeptide repeat GXGVP, where X is an amino acid other than proline, are elastic and share an elastic modulus similar to that of elastin [57]. Polypentapeptides of GVGPG contain β-turns to form a helical β-spiral, a loose water-containing helical structure, with no significant hydrogen bonding between the repeats. The possibility of genetic control of the processing properties, crosslinking, and molecular weight for elastin-like material has established possibilities for their use in tissue engineering. ELPs also display the intrinsic inverse-phase transition properties which is thought to be critical for the self-assembly and alignment of soluble tropoelastin molecules for crosslinking into a native insoluble elastin fibers [58]. Below their characteristic inverse phase transition temperature (T_t), ELPs are soluble in aqueous solution due to formation of an ordered *clathrate shell* of water molecules that shield the hydrophobic domains of elastin chains from each other. When the solution temperature is raised above T_t, the random thermal energy of the component water molecules causes the clathrin hydration shell to break down. The unhydrated ELPs are then able to hydrophobically aggregate and selectively precipitate out of solution to form a highly ordered gel-like solid called a coacervate. The increased entropic disorder of the hydration shell relative to the coacervate provides the thermodynamic driving force for this process [59]. The precise temperature at which the inverse temperature transition takes places depends on the degree of hydrophobicity of the ELP sequence. Inclusion of hydrophilic amino acids in the ELP sequence can interact with the clathrate shell and increase its entropic disorder, thereby decreasing the T_t required for coacervate formation [59]. The possibility of engineering the T_t of ELP solutions at around body temperature lends their utility for in situ self-assembly of cell scaffolds [54, 57, 60].

3
Scaffold Processing

The importance of scaffold geometry in maintaining proper cell–cell geometry and cellular density requires sophisticated processing techniques. A range of scaffold porosities (pore density) and average pore sizes, and hence surface area, are necessary to influence the extent of cell attachment and growth (5 µm for neovascularization; 5–15 µm for fibroblast ingrowth; 20 µm for hepatocyte ingrowth; 40–100 µm for osteoid ingrowth). Pore interconnectivity is also critical to ensure that all cells are within 200 µm from blood supply in order to provide for mass transfer of oxygen, nutrients and waste products [61].

3.1
Alginate

Alginate gels are formed by divalent metal cations which crosslink alginate polymers. The type and density of cations used provides a convenient means of controlling crosslinking density and thus the pore size and mechanical properties of these gels [62, 63]. Since ionic crosslinks are only formed between adjacent G units, crosslinking density can be varied by the choice of G/M block ratio in the bulk material. However, the high variability of alginate G/M ratio, even among those derived from the same sources, has prompted techniques such as enzymatic or chemical treatment of the material to ensure a consistently high G content. Alginate fibers can be fragmented using gamma irradiation of a high-molecular-weight powder (Mn = 170 kDa) to decrease the number-averaged molecular weight (Mn = 0.94 kDa), resulting in an approximate doubling of total void volume [64]. Further options include the introduction of covalent crosslinkers of different sizes and structures, including hydrazines, lysines, and poly-(ethylene glycol)-diamines [61].

Interconnecting macroporous scaffolds with > 90% porosity can typically be processed from alginate gels under mild conditions using lyophilization (freeze-drying) [65, 66] or carbon-dioxide foaming [64] techniques. The freezing regimen can have a critical effect on pore microstructure [67]. Slower freezing kinetics generally lead to the formation of a more homogenous architecture of interconnected and spherical pores. Variable gelation rates of alginate material are also a challenge if uniformly porous structures with complex geometries are the goal. Through an extensive analysis of the effect of different gelation conditions slowing the gelation rate can lead to a more uniform and mechanically stronger hydrogel [68]. Use of lower solubility carbonate or sulfate salts of calcium instead of the more conventional calcium chloride can lead to uniform porosity. Furthermore, use of gelation systems in which the ionic crosslinker is bonded to a carrier [e.g., $CaCO_3$: D-glucono-δ-lactone (GDL)] can allow for controlled release of the crosslinker, thereby providing sufficient time for molding of the alginate suspension into complex shapes before gelation is complete.

3.2
Chitosan

Chitin is usually extracted from crustacean exoskeletons or generated via fungal fermentation processes and subsequent deacetylated under alkaline conditions. The pH-dependent solubility of chitosan allows for convenient processing. Chitosan is first solubilized in dilute acids to protonate the free amino groups. Viscous solutions of highly deacetylated chitosan can be extruded and gelled in a high-pH coagulation solution and then subsequently dried and drawn to form high-strength fibers [69], however it is difficult to weave these fibers into knitted scaffolds. Chitosan matrix manipulation can be accomplished using the inherent electrostatic properties of the molecule. At low ionic strength, the chitosan chains are extended via electrostatic interactions between the acetyl groups. As ionic strength is increased, the inter-chain spacing diminishes and the consequent increase in the junction zone and the stiffness of the matrix results in an increase in average pore size. The pH-dependent charge density and positively charged cationic nature of this material allow it to form insoluble ionic complexes with a wide variety of anionic polysaccharides. Therefore, the formation of such complexes may serve as a convenient means of modifying the surface and bulk properties of chitosan scaffolds.

Porous chitosan structures can also be formed by a freeze-drying process that is similar to that used for alginate. Matthew and others have used this process to fabricate various porous scaffolds with a wide array of pore architectures, including porous membranes, blocks, and tubes [70]. When chitosan–acetic acid solutions are frozen in a glass mold of the desired geometry, ice crystals nucleate from solution and grow along the line of thermal gradients while excluding the chitosan acetate salt from the ice crystal phase. Subsequent ice removal by lyophilization generates a porous material. Mean pore diameters can be controlled within the range 1–250 µm by varying the freezing temperature, the cooling rate, and the rate of ice crystal growth. Chitosan concentration also plays a minor role in determining pore size, with higher concentrations producing smaller pores. Hydration of the lyophilized scaffolds must be performed in a graded ethanol series to prevent the pH-induced changes in crystallinity and associated structural stresses. Any entrapped bubbles are easily removed by brief vacuum treatment. The upper limit for the range of available pore sizes can be increased with the use of an internal bubbling process in which a chitin gel is cast by incorporation of calcium carbonate [71]. Incorporation of calcium carbonate in the chitosan gel reacts with strong acid solution of hydrogen chloride to produce calcium chloride and a large amount of carbon dioxide gas that bubbles throughout the matrix. By varying the amount of calcium carbonate added, gas bubbles of 100–1000 µm can be obtained. Calcium chloride and gas can be readily removed by washing the chitin gel with water and vacuum treatment. As an alternative to scaffold processing, chitosan solutions have been devised that can be combined with cells to form gels in situ at a wound site [72].

3.3
Hyaluronate

Highly purified hyaluronate of > 100 000 Da can be obtained from natural sources (e.g. rooster combs) or artificially produced via microbial fermentation. Inter- and intra-chain esterification of the carboxyl moieties of quaternary salts of hyaluronan can then be readily performed in aprotic solvents using one of a variety of alcohols as the nucleophile. Esterification reduces the solubility of the molecule by reducing the anionic charge of the hyaluronan and increases hydrophobicity to increase viscosity and promote hydrogel formation [31]. These modifications also cause an increase in structural rigidity due to the interaction of hydrophobic groups in organized patches along the chain. The esterified material can then be easily processed by extrusion into fibers or membranes or lyophilized to obtain sponges. Fibers can then be processed into ropes, sleeves, or non-woven articles of various shapes and sizes.

Chemical HA derivatives that have been investigated in detail are the ethyl and benzyl esters of HA, termed Hyaff-7™ and Hyaff-11 [31]. Hyaff-11 is a much less extensively hydrated material than Hyaff-7 (45% vs. 235% increase in weight [73]). Furthermore, it is a soft and readily moldable scaffold, allowing for simple cutting and shaping into appropriate sizes. HA total benzyl ester thread is produced by phase inversion technology. It is then cut, carded and needle-punched in order to obtain a non-woven fleece or a swellable sponge with a uniform structure of open interconnected pores ranging in size from 50 to 340 μm [33].

3.4
Collagen

Collagen is amenable to a wide variety of processing techniques and has been formulated into films, strips, sheets, sponges, beads, discs, and other articles. Collagen can also be electrospun into nanometer-wide fibers [74]. It is readily solubilized in acidic aqueous solutions and can be engineered to exhibit tailor-made properties. One novel collagen scaffold has recently been described as having well-defined and reproducible channel features (from 135 μm to several mm wide) which serve as *artificial vasculature* to increase the mass transport of nutrients and removal of waste products from the construct [75]. Using a controllable layering technique called solid freeform fabrication (SFF), a mold can be designed to precise specifications in order to cast an emulsion of collagen. The mold is subsequently removed by dissolution in ethanol and the resultant solid collagen scaffold is produced by critical-point drying with carbon dioxide such that no residual mold or solvent remains. This processing technique also produces a porous surface due to the removal of ice crystals formed during freezing of the collagen dispersion. Detailed infrared (IR) spectral analysis of the collagen at each phase of the manufacturing process indicates that it maintains its original triple-helical structure and is not denatured.

3.5
Silk

Silkworm silks have been used for centuries in the manufacture of a variety of textiles and articles. Virgin silk from silkworm cocoon fibers must be heat-treated (*degummed*) to remove the associated glue-like proteins called sericins that have been identified as hypoallergenic [76]. Recognition of the superior properties of dragline spider silk in relation to that of the silkworm has prompted investigation into its use as a biomaterial. However, the inability to domesticate spiders and the low yield of material from these organisms has led to attempts at transgenic expression of silk. Unfortunately, the highly repetitive nature of the fibroin peptide sequence usually leads to the synthesis of truncated and insoluble protein in bacterial and yeast expression systems. Successful full-length expression of soluble spider-silk fibroin in the milk of transgenic animals may provide a viable alternative to the use of native fibers [77].

The spinning of reconstituted native silkworm or spider silks has required harsh solvents (ex. lithium bromide or lithium thiocyanate) to solubilize the protein and prevent its premature precipitation by self-assembly and microfibril formation. However, incorporation of a kinase recognition or methionine redox *sterical trigger sites* adjacent to polyalanine sequences has suggested the possibility of biochemical manipulation of fibroin solubility by artificial induction of β-sheet formation [78, 79]. Silk fibers can be artificially wet-spun in aqueous solvents and subjected to post-spinning draw to produce fibers of improved strength (due to an increased number and alignment of crystalline β-sheet regions), but these techniques are insufficient in replicating native fiber properties [80]. Micron-thick native silk fibers ($\sim 5\,\mu m$) can also be spun into much thinner, nanometer-scale, fibers using electrospinning procedures previously devised for collagen [81].

Facile derivatization of the silk fibroin polypeptide is a major advantage when attempting to tailor its material properties to specific uses. Stable amide bond formation between free amines of bioactive peptides and carboxylates of aspartic acid and glutamic acid side chains is possible using conventional chemical coupling chemistry such as via the use of carbodiimide chemistry [82, 83].

3.6
Elastin

The limited employment of elastin as a biomaterial to date is primarily due to difficulties in processing sufficiently pure quantities free of the associated microfibrillar proteins that can lead to an immunological response. This contamination problem is compounded by the strong tendency of the microfibrillar proteins to bind calcium and calcify the associated elastin [84].

A stepwise extraction process from connective tissue has recently been shown to generate highly purified and soluble preparations of elastin, but it is not yet known whether these preparations are sufficiently biocompatible [85].

Bacterial expression of genetically engineered elastin-like polypentapeptides (ELP) may overcome many of the processing problems associated with native elastin. Although traditional cloning is hampered by the instability of the highly repetitive DNA sequence in transformed hosts, iterative strategies for cloning synthetic genes have been devised that allow for precise control over the number and distribution of repeat sequences and total chain length and hence the molecular weight, stereochemistry, and material properties of the ELP [86]. By varying the hydrophobicity of the X amino acid, the so-called guest residue in the VXGVP repeat, one can modify the characteristic inverse phase-transition temperature (T_t) of an ELP solution due to hydrophobic folding and assembly in water. At increased temperatures, the amount of bound water must be decreased causing the polypeptide chains to collapse and form inter- or intramolecule contacts. In particular, a $5 : 3 : 2$ ratio of valine, glycine, and alanine guest residues generate a T_t of 35 °C. This thermal *set point* allows for a number of scaffold engineering possibilities, such as injection of a room-temperature ELP-cell suspension that gels in situ at the physiological temperature of the wound site [57].

Chemical crosslinking of ELPs can facilitate the formation of swellable hydrogels that are nearly completely elastic. Random incorporation of X in the guest residue position of the VXGVP repeat allows crosslinking with tris-succinimidyl aminotriacetate [87] or isoleucine, glutamate, or lysine as the guest residue in GXGVP allows crosslinking with dicumyl peroxide (DCP) radicals or carbodiimide reagents [88] at temperatures above their T_t. These form filamentous structures with reversible swellability and increased tensile modulus. These crosslinked structures can be processed into matrices and fibers that replicate the resilience properties that are conferred by desmosomes in natural elastin.

4
Mechanical Properties

To be used successfully in tissue engineering, it is critical that a biomaterial scaffold temporarily withstand and conduct the loads and stresses that the new tissue will ultimately bear. It is important therefore, to evaluate one or more of the following rheological parameters:

1. Elastic modulus – measured strain in response to a given tensile or compressive stress along the fiber axis;
2. Shear modulus – measured strain in response to a given tensile or compressive stress perpendicular to the fiber axis;

3. Tensile strength – maximum stress that the material can withstand before it breaks; or
4. Maximum strain – Ductility of a material, or total strain exhibited prior to fracture.

Table 2 outlines some of documented mechanical biological parameters for the bulk form of biomaterials discussed in this review. Of the biological polysaccharides, alginate is the most ductile but its compressive strength and elasticity varies considerably with the type of algae from which the material is derived. Alginates with a high G/M ratio (> 0.7) and average G block length (> 15 G residues), such as those from the outer cortex of *Laminaria hyperborea*, are noted for their high mechanical strength [29]. This has been attributed to the egg-box model of cooperative binding whereby divalent cations (the *eggs*) cooperatively bind to a series of negatively charged pockets (the *egg-box*) formed by diaxially linked G block residues. Such cooperative binding in longer G blocks facilitates the formation of extended junctional zones with multiple intra-strand crosslinks.

Significant improvements in rigidity and strength of alginate hydrogels have been attained by optimization of ionic crosslinking systems and gelation conditions [68] or the introduction of covalent crosslinkers of different sizes and structures [62]. Mooney and colleagues have also found that partial oxidation of polyguluronate segments can impart increased rigidity to the molecule [94]. These mechanical improvements are critical for successful use of alginate in tissue engineering since the gels dramatically soften in vivo due to ionic exchange with non-crosslinking ions (e.g., sodium, magnesium) that are present at much higher concentrations. Within the first 15 hours of exposure to physiologically relevant saline conditions, a Ca^{2+} alginate gel can lose more than 60% of its compressive modulus before stabilizing at an equilibrium crosslinking density [63]. This reduction in crosslinking density is accompanied by elevated swelling of the gel and a decrease in viscoelastic behavior that makes the material more susceptible to deformation with functional loading.

Mechanical performance of HA gels are similarly limited. Fiber processing of the total benzyl ethyl derivative Hyaff-11 into non-woven mesh that is pressed during manufacture can lead to a more rigid and stronger gel capable of withstanding higher loads but progressive fiber fraying leads to gradual mechanical failure of the material [90]. A significant decrease in mechanical strength is observed upon hydration of the material.

In comparison to alginate and HA, chitosan is a relatively stiff and rigid bulk material as a result of its high crystallinity. Nonetheless, the very high modulus of elasticity and low strain to failure result in brittle and fragile gels. Improvements in ductility and strength have been achieved using 50/50 blends of bulk chitosan and polyethylene glycol [89]. Hydration and pore formation dramatically improve the flexibility and ductility of material. De-

Table 2 Mechanical properties of naturally derived biomaterials

Biomaterial	Elastic modulus	Shear modulus	Tensile strength	Maximum strain (%)	Refs.
Polysaccharides					
i) MP Ca-alginate [1]	10–12 KPa	9–11 kPa			LeRoux et al. 1999 [63]
ii) Bulk chitosan [2]	166 MPa			8	Kohle et al. 2003 [89]
iii) Porous chitosan [3]	10–15 MPa		30–60 KPa	30–110	Madihally et al. 2000 [70]
iv) Hyaff-11 mesh [4]			0.47 KPa	16.5	Milella et al. 2002 [90]
Fibrous Proteins					
v) Collagen, X-linked [5]	360 MPa*		36 MPa	12–16	Gentleman et al. 2003 [91]
vi) Collagen, X-linked [6]	0.4–0.8 GPa		47–72 Mpa	19	Pins et al. 1997 [39]
vi) Silkworm silk [6]	5–12 GPa		500 MPa	17–18	Perez-Riguiero et al. 2000 [92]
vii) Spider silk [8]	11–13 GPa		740 MPa		Cunniff et al. 1994 [93]
viii) ELP, X-linked gel [9]		2–15 kPa			Trabbic-Wilson et al. 2003 [87]
Human Native Tissues					
ix) Bone	3–30 GPa		60–160 MPa		Yang et al. 2000 [61]
x) Cartilage	1–15 MPa		4–11 MPa		Yang et al. 2000 [61]
xi) Ligament	0.1–0.5 GPa		13–46 MPa		Yang et al. 2000 [61]

[1] Compressive stress of 2% *Macrocystis pyrifera* (MP) alginate gel (MW = 5×10^4, G/M = 0.6); 1.8 mM $CaCl_2$

[2] 93% deactylated chitosan, MW = 6×10^5

[3] Porous hydrogel of 85–90% deactylated chitosan (pore size 1–250 μm)

[4] Pressed, non-woven hyaluronic acid benzyl-ester (Hyaff-11)

[5] 125-μm-diameter fiber extruded from type I collagen dispersion of bovine achilles tendon, carbodiimide crosslinked

[6] Rat-tail collagen dehydrothermally crosslinked and tested after stretching from 0–50%

[7] *Bombyx mori*, naturally produced multi-thread fibers

[8] *N. clavipes*, naturally produced silk fibers

[9] VPGXG, X = Val or Lys coacervate gel, crosslinked with tris-succinimidyl aminotriacetate (125 mg/ml), tested at 37 °C

* Mean tangential modulus calculated from nonlinear stress–strain curve

pending on the pore size and orientation, maximum strains of up to 110% have been reported [68].

The low strength and rigidity of the aforementioned polysaccharides limit their use to soft-tissue applications. Fortunately, the options for tissue engineering are expanded by the use of fibrous proteins, whose normal function is to provide mechanical integrity and stability to biological structures. A growing body of evidence also suggests that fibrous proteins are responsible for the transduction of external mechanical forces to associated cells in a manner that influences the outcome of tissue growth [74]. Until recently, collagen scaffolds have been the biomaterial of choice where a mechanically demanding use is required. For example, type I bovine collagen fibers extruded from bovine tendon have been shown to approximate the mechanical properties of human ligament tissue [91]. Crosslinked collagen sponges have also displayed the ability to remain elastic following long-term application of cyclic stresses that appear critical to the proper functional development of tissues operating in a mechanically dynamic environment [95].

Silks provide the potential for more dramatic improvements to the tensile properties of tissue engineering scaffolds. Dragline spider silks are characterized by their extremely high strength (15% lower than that of high-tensile engineering steel) coupled with much lower relative density (83% lower than steel), excellent elasticity, and resistance to failure in compression.

Complex shear modulus of ELP dramatically increases upon formation of a coacervate [57]. This property is extremely advantageous as it allows for in situ gelation of a functional cell matrix that immediately contributes to the mechanical stability of a wound site. Fine tuning of the mechanical properties may be achieved by altering the concentration of ELP in the injection solution. At 324 mg/ml and an angular frequency of 10 rad/sec, the uncrosslinked ELP coacervate has a complex shear modulus of 80 Pa. Crosslinking methods should further improve this. A crosslinked elastin GVGVP matrix has an elastic modulus approximating that of natural vascular wall (105 N m-2) [96], and recent experiments with the crosslinker tris-succinimidyl aminotriacetate have resulted in gels with a stiffness of 2–15 kPa at 37 °C, a 500-fold increase in stiffness [87]. More natural desmosine crosslinks between adjacent lysine residues can be created by transfecting scaffold cells with lysyl oxidase or increasing the number of lysine residues within the ELP sequence [97]. It is important to note, however, that the mechanical properties of bulk biomaterials are altered by their processing into scaffolds of various pore sizes and pore orientations and further, that these properties will rapidly diminish as a function of implantation time [98].

5
Biodegradability

When used in tissue engineering, the intended fate of a biomaterial scaffold is its eventual permanent replacement by functional host tissue. Ideally the scaffold should degrade at a rate that is inversely proportional to rate of tissue regrowth. Variability of pore structure with biodegradability and cellular growth rate with cell type and microenvironment (Table 3) necessitate careful consideration of the type of scaffold that is used and places high value on materials where processing conditions or polymer derivatization provide a method of controlling the rate of degradation.

In general, hydrogel-forming polysaccharides have a relatively short half-life and their degradation kinetics are inversely related to degree of crystallinity. Highly deacetylated chitosans (> 70%) degrade more slowly in vivo [99]. The derivatization of polymeric side chains with large bulky and hydrophobic substituents can also impart enhanced biological stability to the molecular structure. For example, an increase in benzyl esterification from 75% (Hyaff-11p75) to 100% (Hyaff-11) can result in a tenfold extension in residence time [100].

Chitosan and hyaluronate are degraded by the action of the glycosidic enzymes lysozyme and hyaluronidase, respectively, resulting in oligosaccharide products of variable length. Ultimately, these oligosaccharides are further hydrolyzed to the component monosaccharides, which can be incorporated into glycoproteins or serve as an energy source following conversion to glucose. Total resorption of these implanted materials into physiologically compatible byproducts is a major advantage, and therefore attempts to derivatize these molecules should ensure that toxic byproducts are not produced. For example, spontaneous hydrolysis of the ester substituents of Hyaff-11 has been shown to generate nontoxic free benzyl alcohol in vitro [31, 90] without activation of the complement cascade or opsonization and subsequent recruitment of macrophages.

With respect to degradation kinetics, substantial differences have been observed based on the degree of esterification and the type of ligand introduced on HA esters in vitro [101]. Generally, the rate of ester hydrolysis varies considerably, which can be accounted for by the differing hydration properties of these polymers. Total benzyl esters, Hyaff-11, have a slower degradation rate, which may be attributed to the presence of hydrophobic patches which gives the molecule a more rigid and less mobile conformation, reducing its ability to interact with water for ester hydrolysis. As the degree of esterification decreases, the material becomes increasingly hydrated and soluble and similar to native HA. By contrast, main chain depolymerization occurs at a slower rate due to the action of hyaluronidase, however, polymers with a higher degree of esterification appear to be less susceptible.

In contrast to chitosan and hyaluronate, mammals lack a glycosidase capable of degrading alginate. Instead, the gel is dissolved through ion exchange with non-crosslinking ions in the implant environment, leaving high-molecular-weight uncrosslinked alginate fibers that can persist at the implant site [63]. Partial periodate oxidation of uronate residues [108] or gamma irradiation of material [109] have generated lower-molecular-weight polymers that are more readily excreted from the body.

Fibrous proteins are also somewhat stable in vivo in comparison to some of the polysaccharides, as can be expected from their primary structural role, however they remain susceptible to degradation by proteolytic enzymes. Uncrosslinked collagen can be rapidly degraded by collagenase following implantation but the material can be made increasingly resistant to degradation by manipulation of the crosslinking density. The use of glutaraldehyde is necessary in order to detect any remaining collagen matrix following six weeks implantation in rats [110]. Silk is slowly degraded by proteolysis in vivo despite its tendency to be classified as a nondegradable material [111]. Proteolysis is usually mediated by the foreign-body reaction in a process that favors degradation of the less crystalline portions of the material to peptides that are capable of phagocytosis and metabolic breakdown by macrophages and other foreign body cells. In general, this process is slower than collagen degradation. Silk fibers retain the majority of their tensile strength and modulus at six weeks post-implantation, while crosslinked preparations of collagen lose mechanical integrity in this time frame [106].

To date, the limited in vivo studies with elastin have shown it to be extremely resistant to degradation [107]. Turnover of mature elastin is slow and it can be considered to last for entire lifetimes in the host, despite the presence of endogenous proteases capable of its degradation [112].

6
Biocompatibility and Host Response

Successful use of naturally derived biomaterials requires that they elicit appropriate and beneficial responses from the cells with which they are seeded or from the tissue that is being targeted. These favorable interactions should take place without any potential for harm due to induced cytotoxicity, adverse immune response, or activation of the blood clotting or complement cascades. The type of tissue response invoked by the material is strongly influenced by the site of implantation, the host animal species, and the shape and size of the implant (Table 4). Intrinsic properties of the material such as its chemical reactivity, its mechanism and rate of degradation, and the byproducts of degradation also play a role and can be controlled.

Table 3 Biodegradation of naturally derived biomaterial scaffolds

Biomaterial	Model	Mechanism	Degradation time	Assessment method	Refs.
Polysaccharides					
Alginate, 1–3% Gel	In vitro, physiolocal saline	Dissolution	7 days	63% decrease in compressive moduli	Le Roux et al. 1999 [63]
Hyaluronan films (Hyaff-11; Hyaff-11p75; Hyaff-7)	In vitro, artificial plasma	Ester hydrolysis; hyaluronidase backbone degradation	1–2 weeks (Hyaff-11p75) and 2 months (Hyaff 7 and 11) for total deesterification; 1–4 months for HA backbone degradation	Benzyl alcohol detected by UV/HPLC; Hyaluronan by quantification of GlcNac monomer	Zhong et al. 1994 [101]
Hyaluronan Ester, films	Subcutaneous/ intramuscluar, rat	Hyaluronidase degradation + ester hydrolysis	3–4 months (Hyaff-11) < 1 month (Hyaff-7, 11p75)	Disappearance of material as evidenced by gas chromatography of benzyl alcohol	Benedetti 1998 [102] Campoccia 1996 [103]
Hyaluronan ester (Hyaff-11), plugs	Intraocular, rabbit	Hyaluronidase degradation + ester hydrolysis	5 weeks	Decrease in diameter by ultrasound	Avitabile et al. 2001 [100]
Chitosan, X-linked microspheres	Intramuscular, rat	Lysozyme degradation	> 20 weeks	Weight loss	Mi et al. 2002 [104]
Fibrous proteins					
Collagen, BD90 X-linked[1]	Subcutaneous, rat	Collagenase degradation	> 6 weeks	Maintenance of original implant size; few visible degradation products; light microscopy	Van Wachem et al. 1999 [105]

Table 3 (continued)

Biomaterial	Model	Mechanism	Degradation time	Assessment method	Refs.
Silk, presumed black-braided	Subcutaneous, rat	Foreign body response	6 weeks	55% decrease in tensile strength; 16% decrease in elastic modulus	Greenwald 1994 [106]
Elastin	Subcutaneous, guinea pig	Elastase matrix metalloproteases (MMP-2, MMP-9)	Years??	No observed degradation	Urry et al. 1998 [107]

Table 4 Host response to naturally derived biomaterials

Biomaterial	Model	Observed tissue response	Observed immune response	Onset/ duration	Refs.
A) *Biological polysaccharides*					
Alginate[1], Ba[2+] X-linked	Intrakidney injection, rat		Lack of fibrotic capsule and foreign-body response; slight fibrotic reaction at kidney surface	Analysis at 3 weeks	Klock et al. 1997 [66]
Chitosan, 92% deacetylated porous tubular scaffold	Intraperitoneal & subcutaneous, rat	Collagen deposition and angiogenesis in pore spaces	Early neutrophil accumulation (week 1) but no indication of activation or an inflammatory reaction	1 week (4 weeks)	Van de Vord et al. 2001 [113]
Hyaluronan (Hyaff-11, Hyaff-7, Hyaff-11p75)	Subcutaneous or intramuscular implantation, rat	Cell attachment and spreading with Hyaff-11; cell aggregation and clustering with Hyaff-7, 11p75	Temporary presence of large numbers of foamy macrophages restricted to implantation site, internalize dissolved material	1 week (native HA); 2 weeks (Hyaff-11p75) 1–3 months (Hyaff-11)	Benedetti 1998 [102]; Campocchia 1996 [103]

[1] Highly purified, low-G-content alginate microcapsules (200–300 μm) from *D. potatorum* (0.38 G/C ratio)

[2] Dermal sheep collagen crosslinked at pH 9.0 with 1,4-butanediol diglycidyl ether

[3] Inclusion of (GVGVAP)$_8$ site

Table 4 (continued)

Biomaterial	Model	Observed tissue response	Observed immune response	Onset/duration	Refs.
B) *Fibrous proteins*					
Collagen, BD90 X-linked[2]	Subcutaneous implantation, rat	New collagen formation at week 3	Limited cellular infiltration; small accumulation of mast cells, giant cells, and fibrin networks	2–5 days (6 weeks)	Van Wachem et al. 1999 [105]
Silk, black-braided	Subcutaneous implantation, rabbit		Moderate clustering of histiocytes, lymphocytes and giant cells; significantly thicker capsules than commonly used sutures	30 days (> 120 days)	Setzen et al. 1997 [114]
	Intraarterial implantation, rat		Thrombus encapsulation with platelets, erythrocytes, lymphoscytes	7 days/ (< 28 days)	Dahlke et al. 1980 [115]
$(GVGP)_{216}$ ELP[3], RGD-modified	Subcutaneous implantation, guinea pig	Elastin and collagen production; vascularization	Fine fibrous encapsulation; invasion of macrophages	4 weeks	Urry et al. 1998 [107]

[1] Highly purified, low-G-content alginate microcapsules (200–300 μm) from *D. potatorum* (0.38 G/C ratio)
[2] Dermal sheep collagen crosslinked at pH 9.0 with 1,4-butanediol diglycidyl ether
[3] Inclusion of $(GVGVAP)_8$ site

6.1
Biological Polysaccharides

Of the polysaccharide hydrogels, chitosan and hyaluronate are well tolerated and evoke a minimal foreign-body reaction with no major fibrous encapsulation [101, 116, 117]. Until recently however, in vitro biocompatibility experiments with commercial alginates indicated that they were capable of stimulating macrophages and lymphocytes to release inflammatory cytokines [118]. However, such immunogenicity concerns have been largely overcome as biocompatibility studies with highly purified and crude alginate preparation have shown that many of these observations can be attributed to pyrogenic and immunogenic contaminants in the raw material [66, 119]. Fermentation processes for alginate production will likely ensure reproducibly low levels of endotoxin and mitogens in alginate preparations [120].

There are no species variations in the chemical or physical structure of the HA molecule, which is probably why it is not antigenic and fails to produce a foreign-body reaction [102]. Partial HA esters are likewise biocompatible and exhibit only a transient local response that is normal for the turnover and catabolism of physiological components. No delayed foreign-body response is observed with the material. Unlike the more hydrophobic total benzyl ester Hyaff-11, the more highly hydrated HA esters (Hyaff-7, Hyaff-11p75) are generally unattractive to fibroblasts, which demonstrate little affinity for their surfaces and instead prefer to aggregate with other cells [31]. This lack of cell interaction may be due to the lack of protein adsorption and cell–receptor interactions and may extend to leukocyte interactions. Hyaff-11 failed to activate polarization of neutrophil morphology in vitro [121] or the production of macrophage activation markers nitric oxide and tumor necrosis factor (TNF), in vitro [31], while the more hydrated polymers Hyaff-7 and Hyaff11p75 had moderate effects. Enzyme immunoassay experiments for complement and fibrinolysis activation also confirm the inert nature of HYAFF11 upon exposure to human blood. Moderate activation of the complement pathway, probably at the level of the alternative pathway as evidenced by increased factor Bb and iC3b levels, has been observed with native HA and partially esterified ester Hyaff11p75 [31]. The slight procoagulant nature of native HA and partially esterified esters due to the capacity for binding fibrinogen and inhibiting of fibrinolysis, also indicate that masking the free carboxyl moiety of these materials with an ester group is effective in making the material non-thrombogenic.

In the case of chitosan, biocompatibility occurs despite the onset of an early inflammatory phase that is normally predictive of a robust immune response. Preliminary evidence has shown that the stimulatory effect of chitosan on macrophages and neutrophils is also dependent on the extent of deacetylation [122, 123]. Polymorphonuclear (PMNs) cells and macrophages accumulate rapidly at the implantation site, leading to the formation of granulation tissue in the presence of angiogenesis. This process has been shown to directly activate the recruitment of PMNs through the production of the chemotactic complement factor C5a [123]. Other chemotactic factors, such as prostaglandin E_2 (PGE_2) are detected along with PMNs and macrophages from exudates recovered from the site of chitin administration in dogs. PGE_2 is known to produce arachidonic acid products such a leukotriene-B_4 (LTB_4) which attracts PMNs through the upregulation of endothelial leukocyte adhesion molecules. Isolation of the chemoattractants from chitosan exudates have been shown to promote in vitro PMN migration and recruitment.

VandeVord and colleagues investigated further the biocompatibility of a 92% deacetylated and unseeded sterile, tubular chitosan scaffold (mean pore size = 50 μm) in mice [113]. In agreement with earlier studies, neutrophil infiltration was observed immediately following implantation, but their presence was not prolonged and there was little evidence of an inflammatory response against the material itself since the neutrophils did not appear to actively produce myeloperoxidase and participate in the degradation of the scaffold. The presence of connective tissue matrix within the pore spaces of the implant also increased over time. A highly cellular capsule was also present during the early stages of implantation, but this rapidly decreased in thickness over time, indicating the lack of an adverse foreign-body reaction and rather a chemotactic response to chitosan. The carbohydrate moieties on chitosan may interact with lectin receptors on these cells. Due to the high crystallinity (92% deacetylation), the material maintained its structural integrity for the entire three-month study. Cell recruitment to the biomaterial and tissue ingrowth was limited, probably due to the small size (< 50 μm) of the pores. Despite the presence of leukocytes at the implant site throughout the term of the study, enzyme-linked immunosorbent assay (ELISA) techniques failed to identify any antibodies raised against the material.

6.2
Fibrous Proteins

An extensive body of literature built on years of medical experience with collagen indicates that this protein is biocompatible and non-antigenic [124]. Nevertheless, there are some concerns about the immunogenicity and safety of type I collagen. Adverse reactions have been reported with some local and acute inflammatory response based on prostaglandin E2 production upon in vitro exposure to dermal skin, dependent on the formulation [125]. Since antigenic determinants on the peptide chains of type I collagen have been reported to reside mainly on the telopeptide regions of the molecule, an Atelocollagen® gel has been derived to allay those concerns [126]. Of greater concern is the potential cytotoxicity and pro-calcification effects of carbodiimide, glutaraldehyde, azides, and other crosslinkers that are used to lower collagen antigenicity and increase its stability [127–131]. Glutaraldehyde (GA), the most popular crosslinker, induced apoptosis of human osteoblasts in a dose-dependent manner [132]. In some cases, the crosslinked collagen fibers can be rendered nontoxic by careful washing or subsequent chemical treatment to remove or quench the effects of the crosslinker. Van Luyn and colleagues have performed extensive biocompatibility evaluations of various crosslinked collagen systems, and have identified an epoxy-linked preparation of dermal sheep collagen as a candidate for improved biocompatibility [105, 129, 133]. Use of the crosslinker 1,4-butanediol diglycidyl ether generated a relatively stiff material that was more resistant to enzymatic degradation and induced only a minimal foreign-body response with no evidence of calcification.

Silk has been used less over the last several decades due to observed cases of allergic hypersensitivity [134]. These cases have now been attributed to presence of residual sericin proteins in unprocessed, virgin silk and not silk fibroin itself [111]. This conclusion is supported by the observation that black-braided silk preparations, where sericin in removed and replaced with a wax or silicone coatings, fail to elicit the delayed, type I (IgE-mediated) hypersensitive asthmatic reaction that was occasionally observed in patients several months after exposure to virgin silk [135] and the demonstration that the immunoaffinity of IgE isolated from sera of allergenic patients is directed to the sericin proteins [136].

Studies with a variety of implanted fibroin biomaterials, mostly sutures, have led to varying observations of the degree of foreign-body reaction. This may be due to the differences in implantation site and implant surface area, topography, and size. In general, the inflammatory reaction to black-braided silk is considered moderate with the formation of a fibrous capsule containing giant cells [114]. When implanted within vascular tissue however, black-braided silk has been shown to be acutely thrombogenic. The material binds fibrin and is encapsulated by a dense thrombus [115]. It is eventually resorbed so that several months after implantation only traces of thrombogenicity remain. Subsequent studies have implicated that the initial thrombogenicity of the material can be reduced or eliminated by altering surface properties by eliminating the wax coatings used in black-braided silks [137]. Detailed in vitro studies of the inflammatory potential of degummed fibroin films have suggested that fibroin, especially its hydrophobic patches, can bind and activate fibrinogen and components of alternative complement and humoral immune response pathways [138]. However, the magnitude of these immune responses is similar to that of commonly used synthetic suture materials. Adhesion experiments with macrophages and other components of the cellular immune system demonstrated a more limited response [139]. In actuality, very few studies have directly implicated silk fibroin itself as being responsible for an adverse host response, and modifications to silk processing conditions and composition offer opportunities to make the material more innocuous.

Limited studies performed to date with ELPs indicate that the material is generally well-tolerated but resistant to cell adhesion. Inclusion of an integrin adhesion site (GRGDSP) in the polypeptide results in improved fibroblast adhesion [140]. Subcutaneous injection of arginine-glycine-aspartic acid (RGD)-modified ELP polymer in a guinea pig resulted in a histologically undetectable tissue reaction at two-weeks post-implantation [107]. This material showed regrowth of a natural ECM matrix of collagen and elastin with infiltrating vasculature over the course of two weeks. The lack of a noticeable immune response has been suggested to result from lack of antigenic capacity in the molecule, as evidenced by repeated failures to generate an antibody against this protein [96]. Incorporation of the monocyte chemotactic sequence (GVGVAP), can result in appropriate tissue response to increase degradation time. The apparent inertness of ELP in these studies conflicts those of purified elastin, as it has been shown to undergo severe calcification with hydroxyapatite when implanted in rats [112]. This response is thought to occur by a mechanism mediated by matrix metalloproteinase (MMP-2) as evidenced by reduced calcification in rats administered a MMP inhibitor or when the implant was pretreated with aluminum chloride.

7
Applications for Tissue-Engineering Scaffolds

Do cells grow and proliferate on the biomaterial? Do the cells adhere to the material? Do the cells differentiate appropriately? Does the biomaterial permit tissue ingrowth?

7.1
Connective Tissue – Orthopaedic Use

Requirements: Strength and resistance to mechanical constraints and fatigue. Allowance for stress remodeling. Surface designed for cell-specific interactions. Sterilizability. Good durability for long-term implantation. Good integration with bones and muscles. Ease of fabrication.

To date, the polysaccharide hydrogel scaffolds have received intensive study for their use in the engineering of replacement connective tissues, primarily due to their biochemical similarity with the highly hydrated GAG components of connective tissues.

Cartilage has received particular focus due to its avascular nature, low metabolic requirements, limited capacity for self-repair, and the preponderance of degenerative diseases, such as arthritis, in a rapidly aging demographic. Additionally, the use of three-dimensional scaffolds favors the maintenance of the original chondrocyte phenotype under in vitro conditions. The use of pluripotent stem cells puts more stringent demands on the biomaterial since they require appropriate signals to differentiate into the appropriate cell type. If non-directed, the cells will form a physiologically inferior fibrocartilage which differs significantly in composition from normal cartilage.

The biochemical similarity of polysaccharide hydrogels with GAGs appears to be particularly true of chitosan. Its structural similarities to GAGs are mirrored by its apparent ability to mimic their bio-inductive behavior. The GlcNac residues that are present in many GAGs have been implicated as determinants for specific interactions with growth factors, receptors, and adhesion proteins. This suggests that residual GlcNac in chitosan may also have related bioactivities. Muzzarelli first demonstrated the inductive and stimulatory activity of chitosan while investigating its effects on the rebuilding of connective tissue [117]. Chitosan was observed to organize and accelerate the proliferation of stromal cells surrounding a wound site in the dura matter of cats and it appeared to facilitate the deposition of collagen fibres in reconstructed connective tissue. When used as a matrix for cell encapsulation in macrocapsules, precipitated chitosan facilitates both anchorage-dependent and anchorage-independent cell attachment and spreading of fibroblasts and their response to nerve growth factor by neurite extension [141].

7.1.1
Cartilage

7.1.1.1
Chitosan

Chitosan is popular for mimicking the glycosaminoglycan (GAG)-rich environment of cartilage. Its ability to regulate the chondrocytic phenotype and stimulate chondrogenesis supports its use as scaffolds for repair and reconstruction of cartilage [30]. Lahiji and colleagues found that chitosan films supported the in vitro growth of articular chondrocytes with only a slight decrease in cell viability [142]. Cells maintained a morphological appearance similar to that observed in vivo, as opposed to those grown on plastic which appeared spindly with a fibroblastic, flattened appearance. Molecular analysis of gene expression in these cells indicated that chondrocytes continued to express type II collagen and aggrecan. Lu and colleagues have found that injection of a chitosan solution into the knee articular cavity of rats caused a significant increase in chondrocyte density [143]. Thus chitosan can serve as a matrix to support the attachment, proliferation, and maintenance of extracellular phenotype. The stiffness of chitosan relative to other polysaccharide hydrogels has been suggested as a factor in its ability to maintain a sufficient proliferative phenotype with delayed type II collagen deposition and differentiation until an adequate population of chondrocytes can be established in the scaffold [23]. The use of chitosan also prevents dedifferentiation and production of an inferior type I collagen matrix since it subsequently fails to produce long-lasting hyaline cartilage.

7.1.1.2
Alginate

Alginate has been intensively studied in vivo for use in cartilage engineering due to its excellent bio-inductive capacity on chondrocytes. Alginate implants have been shown to be capable of supporting chondrocyte viability and production of a cartilage-like extracellular matrix as early as four weeks after implantation [144]. In contrast to growth on chitosan scaffolds, chondrocytes stop proliferating and tend to more readily produce type II collagen and assume a differentiated phenotype with rounded morphology when grown on an alginate substrate [23]. Cell aggregation and differentiation on alginate matrices have been attributed to the high hydrophilicity that minimizes cell and protein adhesion and allows cell–cell interactions to predominate and rapidly establish functional tissue. The rapid switch from a proliferative to a differentiated phenotype could be problematic if a threshold population of cells has not yet been established, but this potential disadvantage can be largely offset by the efficient, rapid, and high-density seeding that is possible when using a dry alginate scaffold with > 90% porosity [145, 146]. The efficient seeding properties of alginate are facilitated by pore tortuosity which entraps the cells and highly hydrophilic matrix which wicks up the cells when dry, resulting in homogenous distribution throughout the scaffold. The property of high-density cell seeding is desirable for rapid regeneration of ECM and secretion of tissue-regenerating growth factors. Unfortunately, alginate implants fall well short of matching the mechanical properties of native cartilage tissue in mechanically demanding requirements. Therefore, its application will likely be useful to creation of complex 3-D shapes, for use in reconstructing facial structures (ears, nose bridge, etc.) where auricular cartilage is important for maintain proper dimensions, rather than mechanical integrity [147].

7.1.1.3
Hyaluronate

The high HA content of cartilage and the ability of HA to form a highly hydrated polymer matrix with peculiar viscoelastic properties make it a natural candidate for the repair of damaged cartilage tissue. However, the highly hydrated nature of most hyaluronates prevents cell adhesion to the material so that most chondrocytes are lost by the application of shear following implantation. Recently, a thiol-modified HA has been developed [3,3'-dithiobis(propanoic dihydrazide)]-HA which can be modified with RGD and other peptides and chemically crosslinked with PEG diacrylate [148]. This

Table 5 Tissue engineering with biopolymer-based biomaterial scaffolds

Targeted tissue	Biomaterial scaffold	Cell type seeded	In vitro/ in vivo response	Refs.
Bone	Alginate, 1% gel with BMP-2 oligopeptide	Rat primary calvarial osteoblasts	In vivo rat model, subcutaneous: ectopic bone formation in hydrogel pores; zones of woven bone at 3 weeks post-implantation; calcified trabecular bone at 8 weeks	Suzuki et al. 2000 [150]
	Hyaluronan (Hyaff-11)	Rat bone marrow stromal cells	In vivo rat model: acceleration of bone mineralisation (40 vs. 160 days) when supplemented with bFGF	Lisignoli et al. 2002 [151]
	Collagen sponge, type I bovine, alkaline treated	Acellular	In vivo rat model: Osteogenesis and bone ingrowth according to fibril orientation; cancellous bone and bone marrow formation in treated defects (40 days); cortical bone formation in control	Rocha et al. 2002 [152]
Ligament	Collagen gel	Human bone-marrow stromal cells (hBMSC)	In vitro ligament fibroblast differentiation in response to cyclic mechanical stress	Altman et al. 2002 [153]
	Silk fibroin fiber (RGD-coupled)	Human osteoblast-like cells (Saos-2)	In vitro induction of bone formation, bone-specific gene expression	Sofia et al. 2001 [82]
	Silk fibroin fiber (woven cord matrices)	Human anterior cruciate ligament (ACL) fibroblasts	In vitro ligament-specific development, ECM formation	Altman et al. 2002 [154]

Table 5 (continued)

Targeted tissue	Biomaterial scaffold	Cell Type seeded	In vitro/ in vivo response	Refs.
Cartilage	Alginate, 2% Ca^{2+} X-linked gel	Bovine articular chondrocytes	In vivo mouse model, subcutaneous implantation: progressive cartilage formation: multiple islands of cartilage at 6 weeks; solid cartilage implant at 12 weeks; cells maintain rounded morphology; increase in collagen and proteoglycan content	Chang et al. 2001 [144]
	Chitosan	Acellular	In vivo rat model; knee injection; increased articular cartilage chondrocyte density at 6 weeks; decrease in epiphyseal cartilage thickness	
	Hyaluronan (Hyaff-11)	Human articular chondrocytes	In vitro: upregulation of chondrocytes specific transcripts (type II & X collagen; Sox9, aggrecan)	Girotto et al. 2003 [155]
		Autologous rabbit chondrocytes	In vivo rabbit model: improved healing of full-thickness defects in femoral condyle; GAG-rich hyaline matrix	Grigolo et al. 2001 [149]
	Collagen, type I bovine matrix (100–300 μm pore size)	Rabbit/bovine chondrocytes	In vitro: upregulation of some chondrocyte specific transcripts (aggrecan); proteoglycan synthesis; maintenance of spherical cell morphology	Saldanha et al. 2000 [156]; Schuman et al. 1995 [157]
	Collagen, type II porcine matrix, UV crosslinked, 90 μm pore size (Chondrocell)	Canine chondrocytes	In vitro: maintenance of spherical cell morphology (60% of cells at 21 days); upregulation of type II collagen GAG content/DNA at 2 weeks In vivo adult canine model: chondral defect in knee joint; type II seeded matrices showed increased levels of hyaline cartilage	Nehrer et al. 1997 [158]; Nehrer et al. 1998 [159]

Table 5 (continued)

Targeted tissue	Biomaterial scaffold	Cell Type seeded	In vitro/ in vivo response	Refs.
Cartilage	ELP coacervate gel	Pig femoral condyle chondrocytes (in situ gelation)	In vitro: maintain rounded morphology, synthesis of collagen and sulfated GAG-rich ECM	Betre et al. 2002 [57]
Blood vessels	Collagen small diameter vascular graft*	Acellular	In vivo rabbit arterial bypass model: smooth muscle cellularization and endothelializatoin in 3 months (actin and CD31 staining); withstood 930 mm Hg burst pressure; 90-day patency; response to vasoactive agents (norepinephrine, bradykinin)	Huynh et al. 1999 [160]
Smooth muscle	Collagen, bovine type I sponge, GA X-linked	Rat aorta smooth muscle cells	In vitro culturing under cyclic stress (7% amplitude; 0.5 Hz frequency for 20 weeks): Upregulation of elastin synthesis (Verhoeff's stain); seeded cells display contractile phenotype (myofilament bundles; contractile apparatus)	Kim & Mooney 2000 [95]; Kim et al. 1999 [4]

* 4 mm diameter tube of pig submucosal intestinal collagen layer deposited with type I bovine collagen fibrils and carbodiimide cross-linked

facilitates the adhesion and spreading of fibroblasts both in vitro and in vivo following in situ gelation, but did little to promote tissue ingrowth. Grigolo and others have used the total benzyl ester hyaluronan derivative, Hyaff-11, as a carrier for delivery of chondrocytes to a wound site in rabbit cartilage [149]. The more hydrophobic character of this material enabled autologous chondrocytes to strongly adhere and promote regeneration of appropriate hyaline cartilage instead of inferior fibrocartilage, despite the risk of cell detachment by joint movement during the healing process. Analysis of gene expression has demonstrated concurrent upregulation of chondrocyte-specific genes such as collagen type II, type X, and Sox9 following transfer of these cells from a monolayer culture to a Hyaff-11 scaffold [149]. This establishes that Hyaff-11 is able to stimulate chondrocytes to produce a characteristic collagen and GAG-rich extracellular matrix and maintain other aspects of the appropriate chondrocytic phenotype despite having dedifferentiated to a fibroblast-like phenotype in monolayer culture. Therefore, Hyaff-11 provides an excellent support matrix for in vitro chondrocyte culture, generating a functional cartilage construct for implantation. As with alginate, however, the limited mechanical robustness of these scaffolds needs to be addressed.

7.1.1.4
Collagen

In cartilage engineering, collagen has been investigated for some time with early in vivo studies indicating that passaged chondrocytes seeded into collagen scaffolds and directly implanted can improve healing of chondral defects in a canine model [159]. However, the reparative tissue was to shown to consist primarily of unwanted fibroblasts and the focus of using this material turned to control of cell expression and mechanical properties during the in vitro chondrocyte culturing process. To that end, chondrocytes embedded and cultured in a type I collagen hydrogel gradually proliferate and maintain a hyaline cartilage phenotype for up to four weeks, as determined by chondroitin-6-sulfate production, matrix gene expression, or other analytical methods [126, 156]. Moreover, the effects of mechanical compression on in vitro extracellular matrix synthesis by chondrocyte-seeded type II collagen scaffolds has been shown to more closely mirror biosynthesis and cell behavior of native articular cartilage than type I collagen [158, 161]. By analyzing the incorporation rates of radio-labeled proline and sulfur, it was shown that collagen and proteoglycan are expressed to a higher extent, probably because type II collagen is a closer analog to the ECM of articular cartilage [161]. However, any observed increase in reparative cartilage tissue with type II versus type I collagen was found to be statistically insignificant in a 15-week study of full-thickness chondral defect repair in a canine model [159].

7.1.1.5
Elastin

Betre and colleagues found that chondrocytes in a 3-D ELP coacervate assumed a rounded morphology after 15 days in vitro culture and stained positive for GAGs and collagen, indicating the presence of a functional ECM [57]. Biochemical analysis of chondrocyte-specific sulfated GAGs and collagen components showed an increased accumulation with culture that was independent of cell proliferation, indicating that the material was capable of inducing ECM synthesis reminiscent of native cartilage tissue.

7.1.2
Ligament/Tendon

7.1.2.1
Collagen

The importance of a mechanically dynamic environment during the culturing process has also been demonstrated in attempts to engineer functional ligament–collagen tissue constructs [163]. Application of mechanical stimuli to adhered human stem cells in three-dimensional collagen gel scaffolds has been shown to facilitate ligament-specific cell differentiation and orientation of cells in the direction of loading. Complete tissue ingrowth in vitro allows the implant to attain cellularized ligament *neotissue* into the wound site, thereby minimizing a foreign-body reaction to the collagen scaffold component. Unfortunately, the mechanical integrity of collagen gels and their relatively rapid biodegradation are seen as inadequate for implantation in mechanically demanding environments such as the knee joint.

7.1.2.2
Silk

The superior tensile strength and adequate biocompatibility of silk biomaterials have prompted their use for the engineering of tissues that have stringent requirements for mechanical robustness and integrity, such as the anterior cruciate ligament (ACL). Weaving techniques for the fabrication of millimeter-thick, wire-rope silk cords have been described which mimic the stress–strain properties of native human ACL [76]. These cords are comprised of thousands of silk fibers in a defined higher-order geometry (matrix of 30 fibers × 6 bundles × 3 strands × 6 cords). Human bone-marrow stromal cells (BMSC), mesenchymal adult stem cells, proliferated on these cords, formed cell sheets within 14 days in vitro culture, all with negligible biodegradation of the material. The six-cord matrix offered maximized surface area for cell seeding, adequate filament gaps for tissue ingrowth (5–60 µm), and proper diameter for in vivo insertion into bone tunnels while minimizing mass-transfer limitations. BMSCs expressed appropriate levels of ligament-specific markers (collagen types I and III, tenascin-C), indicating the maintenance of a ligament fibroblast phenotype. Studies have yet to be performed to assess the effects of applied mechanical stimuli on cellularized scaffold, but they would be expected to further approximate authentic ligament tissue. If autologous cells are available for seeding, implantation of the cellularized scaffold could result in integration of an immediately functional *neotissue* that elicits a minimal host response.

7.1.3
Bone

7.1.3.1
Hyaluronan

Despite their weak mechanical properties, polysaccharide hydrogels have demonstrated applicability as osteoprogenitor cell scaffolds for the repair of bone defects in non-weight-bearing areas, such as the craniofacial region. In both rat and rabbit models containing a radial osteochondral defect, implantation of a Hyaff-11 scaffold seeded with bone-marrow stromal cells led to osteoblast differentiation followed by mineralization and regeneration of tissue in the defect [151, 164]. In the rabbit model, the HA scaffold was completely degraded within four months of implantation. Preincubation of the scaffold with basic fibroblast growth factor (bFGF) resulted in a more rapid onset in the production of mineralized matrix. bFGF appeared to promote the events that lead to mineralization, stimulating both osteoblast proliferation and alkaline phosphatase activity. In addition to inducing more rapid mineralization processes, the scaffold was also more homogenous, with matrix reaching the interior of the scaffold.

7.1.3.2
Alginate

Similarly, osteoblasts have also promoted bone mineralization when encapsulated in alginate implants [124, 165]. The requirement for osteoblast adhesion necessitated the use the coupling of alginate matrices with the tripeptide cell-adhesion motif arginine-glycine-aspartic acid (RGD). RGD motifs provide a ligand for integrin receptors which enable cell adhesion to the normally nonadhesive alginate matrix [166, 167]. Alternatively, covalent attachment of an oligopeptide derived from the growth factor bone morphogenic protein-2 (BMP-2) could promote calcification by stimulating the migration of osteoblasts into an alginate scaffold [150]. Mooney and colleagues have also derived semisynthetic injectable poly(aldehyde guluronate) alginate hydrogels that are covalently crosslinked with hydrazine to improve the mechanical performance in vivo [40]. When seeded with calvarial osteoblasts, implanted subcutaneously, and crosslinked in situ, these implants form ectopic calcified tissue within nine weeks, holding promise for their use in reconstructive surgery where moderate loads must be withstood.

In most scenarios, scaffolds for the engineering or repair of bone must concurrently provide both mechanical support and induction of bone mineralization. Three-dimensional cell–cell interaction and cellular density also appear to be crucial for proper cellular differentiation that leads to the formation of mature bony tissues [168]. As such, these applications require complex geometries that retain mechanical strength, particularly performance under compression, where biomaterials such as fibrous proteins excel. Since type I collagen is the primary constituent of bone, it has logically been used in many attempts to engineer this tissue type. In most cases, however, research has focused on the use of acellular collagen implants to promote tissue ingrowth and prevent dystrophic calcification which can lead to functional failure of the cellularized matrices when implanted in bone [152]. Collagen sponge scaffolds have been most successful in promoting osteogenesis when loaded with recombinant bone morphogenic protein (rhBMP-2) and injected in non-load-bearing applications such as periodontal implants [169].

The excellent performance of silk fibroin under compressive stress has motivated attempts to exploit this property by directing bone development on silk matrices. Detailed in vitro studies have demonstrated that silk fibroin shows promise in directing the growth and mineralization of bone, particularly when covalently coupled to RGD peptides [82]. This functional group promotes integrin-mediated attachment of osteoblasts leading to the expression of bone matrix specific such as alkaline phosphatase, $\alpha 1(I)$ procollagen and osteocalcin. Calcein staining and calcium quantification further demonstrated that the adhered cells were capable of manufacturing mineralized matrix. Human bone-marrow stromal cells have also been show to adhere and proliferate to high cell densities when seeded on RGD-modified silk fibers [83]. While the osteo-inductive capabilities of silk derivatives need to be confirmed by in vivo experiments, the positive results generated thus far are encouraging for those wishing to exploit the impressive mechanical properties of silk biomaterials.

7.2
Bioartificial Organs and Soft Tissue Implants

Requirements: Same mechanical properties as the surrounding tissues. No chronic response to toxic leachables. Integration with natural tissues without fibrous reaction. Porosity. Creep resistance.

7.2.1
Alginate

Fragile polysaccharide hydrogel materials such as alginate have demonstrated potential for use in non-load-bearing applications such as repair or reconstruction of soft tissues. When seeded within a macroporous alginate scaffold, progenitor cells of rat heart muscle have been shown to aggregate into islets of cells that fill the available 100-μm pore space and express appropriate tissue-specific genes [146]. Not only does immunostaining of cardiomyocytes demonstrate that contractile proteins are expressed but the spontaneous physical contractions of these cells indicate that they are functional as well. Seeding hepatocytes onto a similar sponge-like alginate scaffold generated cell aggregates capable of secreting albumin and urea and detoxification enzymes (e.g., cytochrome P-450), indicating that their functional status as components of liver tissue was maintained and the mass transfer of nutrients and waste products were not serious limitations [71, 145]. As has been demonstrated with other cell types, the hepatocytes failed to proliferate on the alginate material but maintained a stable population of viable cells over a period of two weeks. Injection of an alginate cell suspension and gelation post-injection has also been demonstrated to improve maintenance of scaffold ultrastructure and alginate stiffness in vivo when compared to pre-gelled scaffolds, particularly when seeded with syngeneic fibroblasts [171].

Most soft-tissue engineering applications will require the introduction of large cellularized scaffolds whose viability would undoubtedly be constrained by mass-transfer limitations. Recently, large cardiomyocyte-seeded alginate scaffolds have been modified by incorporating PLGA microspheres which release the angiogenic growth factor bFGF [172]. Sustained elution of bFGF enhanced the vascularization of the tissue scaffold and promoted long-term growth of the tissue. Furthermore, when implanted in the peritoneal cavity of a rat, the scaffold facilitated vascular ingrowth. Optimization of approaches such as these will be critical for the successful use of biomaterial scaffolds for engineering of large tissues and whole organs.

7.3
Elastic Tissues (Blood Vessels, Smooth Muscle)

Requirements: Wear and tear resistance; fatigue resistance; lubricity; lack of thrombus or embolism formation and chronic inflammatory response; biostability; sterilizability; porosity for tissue ingrowth; mechanical integrity.

7.3.1
Collagen

Blood-vessel engineering has been attempted in which an acellular bovine type I collagen lattice derived from a small-intestine submucosal layer and crosslinked with a water-soluble carobodiimide is able to integrate into the host tissue and provide a scaffold for remodeling into a functional blood vessel [160]. The use of collagen as a material that promotes tissue ingrowth for vascular remodeling avoids extensive pre-surgical culture time to generate the autologous cellularized scaffolds and the challenges of producing a cultured vessel with the required mechanical strength. When implanted into a rabbit as a vascular graft, remodeling by vascular smooth muscle and endothelial cells was observed within three months, and good patency was observed in response to physiological blood pressure (> 900 mm Hg). Grafts also appeared nonthrombogenic and functional as neovessels by the contractile response to hormonal agonists. Collagen has also been investigated for engineering of smooth muscle for cardiovascular, gastrointestinal and urinary bladder applications. In these scenarios, the material should display elastic behavior in order to withstand repetitive cycles of stress and strain. Mooney and colleagues have found that smooth muscle cells cultured under cyclic strain conditions for a 20-week period have a phenotype which is reminiscent of smooth muscle [95]. This indicates that collagen can exhibit elastic-like behavior that is effective in transducing mechanical signals to the cell so that it can adapt to a dynamic environment.

7.3.2
Elastin

ELPs have been investigated for the engineering of soft tissues where elastic modulus and viscoelasticity are important for transmitting tensional or compressional forces. For example, the enhancement of urethral compression by adding tissue bulk for support around a bladder sphincter muscle has been suggested as a possible treatment of urinary incontinence. Indeed, simulated emptying and filling of an RGD-modified ELP matrix has been found to promote the outgrowth of urothelial cells and their elaboration of an ECM [173]. The replacement or restoration of diseased or damaged vertebral discs is another possible use for elastin due to the dynamic loads that these tissues experience. The response of injected elastin to the compression and tension could result in remodeling of the tissue while also providing the viscoelasticity required for functionality. Several studies have suggested that the viscoelastic properties of elastic proteins can sense mechanical forces and transduce mechanical free energy from these forces into biochemical remodeling processes, particularly synthesis of extracellular matrix molecules to help withstand the forces to which the cell is being subjected [96].

8
Future Prospects and Considerations

The need for alternative solutions to meet the demand for replacement organs and tissue parts will continue to drive advances in tissue-engineering biomaterials for the future. When compared with the widely utilized artificial polymers, biologically derived polymers offer several advantages for use in tissue-engineering scaffolds. The biopolymers presented in this review provide tissue engineers with a range of mechanical properties and biodegradation rates which can be appropriately matched with a corresponding tissue remodeling rate. The excellent biocompatibility of polysaccharide hydrogels, such as alginate and chitosan, are suited for use in soft-tissue applications or even non-load-bearing tissue types such as bone or cartilage. Stronger and more slowly degradable materials such as silk and elastin have demonstrated applicability for the regeneration of vascular, ligament and cartilaginous tissues that require resiliency and durability.

It is increasingly recognized that the similarity of these natural biomaterials to ECM components is invaluable for providing a familiar context in which seeded cells can proliferate and differentiate to form desired tissue types. Transplanted cell scaffolds must not only induce integration of the defect but they must also retain the capacity to induce and maintain appropriate growth as the body repairs and matures. Recent breakthroughs have been documented in which co-cultured osteoblasts and chondrocytes were seeded simultaneously on high-density, RGD-modified alginate scaffolds and implanted subcutaneously in rats [174]. The alginate scaffold was not further substituted with exogenous growth factors, yet induced the organization of a structure which closely resembled the epiphyseal bone growth plate, a structure which is responsible for bone elongation during normal development. The neo-growth plate resulted in the development of bony tissues with increased in mass, bone mineralization, type II collagen deposition, and cartilage content with time. Thus, introduction of a small number of different precursor cells and precisely guiding their spatial orientation, distribution, and growth rates of individual populations of cells in vivo, could allow for the introduction of tissue organization centers with the self-regenerating properties of normal tissues.

Artificial and biologically derived polymers and tissue-engineering scaffolds share common hurdles before they become viable alternatives in the clinic. Perhaps the biggest hurdle is the sourcing of appropriate autologous or stem cells for seeding the scaffolds. However, the principal disadvantage of biologically derived polymers in relation to artificial types lies in the development of reproducible production methods. Their structural complexity often renders modification and purification difficult and increases the expense of these materials. Difficulty in maintaining the reproducibility of biopolymer scaffolds due to batch-to-batch and source-to-source variations during manufacture also impose significant regulatory hurdles. Chemical engineering techniques have enabled the formation of adjustable properties of synthetic biomaterials. Recombinant DNA technology is supplanting conventional synthetic approaches to produce component building blocks or homopolymers and block copolymers with good control of polymer length, primary sequence, composition, stereochemistry and structure and with time can be expected to provide improved options in the production of controlled biopolymer products.

Concerns about immunogenicity and disease transmission from animal-derived products such as type I collagen persist, particularly in light of prion-mediated diseases such as bovine spongiform encephalopathy, which are indestructible by conventional sterilization methods. Inflammatory potential from endotoxins must also be considered, particularly since biopolymers are difficult to sterilize and biomaterial-related infections are resistant to host defences and antibiotics. Production of synthetic silk polymers and ELPs do not require isolation from native sources and the production of artificial synthetic human collagens are underway using genetic expression systems [175]. For example, type I human procollagen has been shown to assemble when heterologously co-expressed in yeast with post-translational modifier, prolyl hydroxylase [176].

To tailor scaffolds properly a large toolbox of proteins and oligopeptide units increases the possibility of engineering synthetic silks or elastins with precisely defined functionality. Thus, several groups have initiated work on synthetic silks in which block copolymers based on consensus repeat sequences derived from the alanine-rich hydrophobic and glycine-rich hydrophilic domains of the spridroin 1 and 2 proteins of native dragline spider silk [42, 175]. For example, synthetic genes designed from the consensus peptide sequence of the spider *Nephila clavipes* have been successfully expressed in the methyltropic yeast host, *Pichia pastoris* [177, 178]. Significant progress towards the synthesis of alternative silks have been described [179]. The wealth of structurally and functionally diverse silks among spider and silkworm species (major and minor ampullate silks, flagilliform silks, cocoon silks) remains to be investigated in sufficient detail. These silks are a potential trove of sequence motifs and their contribution to the mechanical properties of each silk type need to be dissected. Similarly, little tissue engineering work has been performed with non-type-I collagens, despite the 20 or more collagen types, each of which is presumed to have been optimized for specific functions. In addition to the tailored design of polymer sequences, the advantages of making semi-synthetic derivatives of biopolymers in order to modify or further elaborate on their properties is being pursued. For example, the esterified hyaluronans, silk conjugates, and novel crosslinked collagens provide improvements in stability or bioactivity. In addition, improvements in scaffold design and manufacture are required to provide scaffolds with sufficient heterogeneity in porosity, texture, and polymer alignment for cell and tissue formation.

Ultimately, the exploitation of biopolymer-based matrices for biomaterials and tissue engineering can be expected to have major impact on the quality of tissue outcomes, due to the natural match in chemistry, structure and functions to native ECMs. The challenges that remain provide many exciting and fruitful fundamental and applied investigations into the control of polymer structure, modification to enhance cell-specific interactions, and increased complexity in material formulation to facilitate the formation of more-complex tissue structures.

References

1. Langer R, Vacanti JP (1993) Tissue engineering. Science 260:920
2. Vacanti JP, Langer R (1999) Tissue engineering: the design and fabrication of living replacement devices for surgical reconstruction and implantation. The Lancet 354:(Suppl)32–34
3. Yoshimoto H, Shin YM, Terai H, Vacanti JP (2003) A biodegradable nanofiber scaffold by electrospinning and its potential for bone tissue engineering. Biomaterials 24(12):2077–2082
4. Kim BS, Mooney DJ (1998) Development of biocompatible synthetic extracellular matrices for tissue engineering. Trends Biotechnol 16:224–230
5. Boontheekul T, Mooney DJ (2003) Protein-based signaling systems in tissue engineering. Curr Opin Biotechnol 14:559–565
6. Angelova N, Hunkeler D (1999) Rationalizing the design of polymeric biomaterials. Trends Biotechnol 17:409–4521
7. Mauck RL, Nicoll SB, Seyhan SL, Ateshian GA, Hung CT (2003) Synergistic action of growth factors and dynamic loading for articular cartilage tissue engineering. Tissue Eng 9(4):597–611
8. Mauck RL, Soltz MA, Wang CC, Wong DD, Chao PH, Valhmu WB, Hung CT, Ateshian GA (2000) Functional tissue engineering of articular cartilage through dynamic loading of chondrocyte-seeded agarose gels. J Biomech Eng 122(3):252–260
9. Gomes ME, Ribeiro AS, Malayfaya PB, Reis RL, Cunha AM (2001) A new approach based on injection molding to produce biodegradable starch-based polymeric scaffolds: morphology, mechanical, and degradation behaviour. Biomaterials 22(9):883–889
10. Tate MC, Shear DA, Hoffman SW, Stein DG, LaPlaca MC (2001) Biocompatibility of methylcellulose-based constructs designed for intracerebral gelation following experimental traumatic brain injury. Biomaterials 22(10):1113–1123
11. Ehrenfreund-Kleinman T, Domb AJ, Golenser J (2003) Polysaccharide scaffolds prepared by crosslinking of polysaccharides with chitosan or proteins for cell growth. J Bioact Compat Polym 18(5):323–338
12. Matsuda T, Magoshi T (2002) Preparation of vinylated polysaccharides and photofabrication of tubular scaffolds as potential use in tissue engineering. Biomacromolecules 3(5):942–950
13. Pieper JS, Hafmans T, van Wachem PB, van Luyn MJ, Brouwer LA, Veerkamp JH, van Kuppevelt TH (2002) Loading of collagen-heparan sulfate matrices with bFGF promotes angiogenesis and tissue generation in rats. J Biomed Mater Res 62(2):185–194

14. Pieper JS, van Wachem PB, van Luyn MJA, Brouwer LA, Hafmans T, Veerkamp JH, van Kuppevelt TH (2000) Attachment of glycosaminoglycans to collagenous matrices modulates the tissue response in rats. Biomaterials 21(16):1689–1699

15. Pieper JS, Hafmans T, Veerkamp JH, van Kuppevelt TH (2000) Development of tailor-made collagen-glycosaminoglycan matrices: EDC/NHS crosslinking, and ultrastructural aspects. Biomaterials 21(6):581–593

16. Passaretti D, Silverman RP, Huang W, Kirchoff CH, Ashiku S, Randolph MA (2001) Cultured chondrocytes produce injectable tissue-engineered cartilage in hydrogel polymer. Tissue Eng 7(6):805–815

17. Ye Q, Zund G, Benedikt P, Jockenhoevel S, Hoerstup SP, Sakyama S, Hubbell JA, Turina M (2000) Fibrin gel as a three-dimensional matrix in cardiovascular tissue engineering. Eur J Cardiothor Surg 17(5):587–591

18. Underwood S, Afoke A, Brown RA, MacLeod AJ, Shamlou PA, Dunnill P (2001) Wet extrusion of fibronectin–fibrinogen cables for application in tissue engineering. Biotechnol Bioeng 73(4):295–305

19. Novikov LN, Novikova LN, Mosahebi A, Wiberg M, Terenghi G, Kellerth JO (2002) A novel biodegradable implant for neuronal rescue and regeneration after spinal cord injury. Biomaterials 23(16):3369–3376

20. Sodian R, Sperling JS, Martin DP, Egozy A, Stock U, Mayer JE, Vacanti JP (2000) Fabrication of a trileaflet heart valve scaffold from a polyhydroxyalkanoate biopolyester for use in tissue engineering. Tissue Eng 6(2):183–188

21. Stock UA, Sakamoto T, Hatsuoka S, Martin DP, Nagashima M, Moran AM, Moses MA, Khalil PN, Schoen FJ, Vacanti JP, Mayer JR (2000) Patch augmentation of the pulmonary artery with bioabsorbable polymers and autologous cell seeding. J Thorac Cardiovasc Surg 120(6):1158–1167

22. Stock UA, Nagashima M, Khalil PN, Nollert GD, Herden T, Sperling JS, Moran A, Lien J, Martin DP, Schoen FJ, Vacanti JP, Mayer JE (2000) Tissue engineered valve conduits in pulmonary circulation. J Thorac Cardiovasc Surg 119(4Pt1):732–740

23. Drury JL, Mooney DJ (2003) Hydrogels for tissue engineering: scaffold design variables and applications. Biomaterials 24(24):4337–4351

24. Grassl ED, Oegema TR, Tranquillo RT (2002) Fibrin as an alternative biopolymer to type-I collagen for the fabrication of a media equivalent. J Biomed Mater Res 60(4):607–612

25. Meinhart J, Fussenegger M, Hobling W (1999) Stabilization of fibrin-chondrocyte constructs for cartilage reconstruction. Ann Plas Surg 42(6)

26. Madison LL, Huisman GW (1999) Metabolic engineering of Poly(3-hydroxyalkanoates): From DNA to Plastic. Microbiol Molec Biol Rev 63(1):21–53

27. Martin DP, Williams SF (2003) Medical applications of poly-4-hydroxybutyrate: a strong flexible absorbable biomaterial. Biochem Eng J 16:97–105

28. Williams SF, Martin DP, Horowitz DM, Peoples OP (1999) PHA Applications: addressing the price performance issue I. Tissue Engineering. Int J Biol Macromol 25:111–121

29. Smidsrod O, Skjak-Braek G (1990) Alginate as an immobilization matrix for cells. Trends Biotechnol 8:71–78

30. Suh JKF, Matthew HWT (2000) Application of chitosan-based polysaccharide biomaterials in cartilage tissue engineering. Biomaterials 21:2589–2598

31. Campoccia D, Doherty P, Radice M, Brun P, Abatangelo G, Williams DF (1998) Semisynthetic resorbable materials from hyaluronan esterification. Biomaterials 19:2101–2127

32. Laurent TC, Fraser JR (1994) Hyaluronan. FASEB J 6(7):2397–2404
33. Gamini A, Paoletti S, Toffanin R, Micali F, Michielin L, Bevilacqua C (2002) Structural investigations of cross-linked hyaluronan. Biomaterials 23:1161–1167
34. Bulpitt P, Aeschlimann D (1999) New strategy for chemical modification of hyaluronic acid: preparation of functionalized derivatives and their use in the formation of novel biocompatible hydrogels. J Biomed Mater Res 47(2):152–169
35. Prestwich GD, Marecak DM, Marecek JF, Vercruysse KP, Ziebell MR (1998) Controlled chemical modification of hyaluronic acid: synthesis, applications, and biodegradation of hydrazide derivatives. J Control Release 53(1–3):93–103
36. Baier Leach J, Bivens KA, Patrick CW, Schmidt CE (2003) Photocrosslinked hyaluronic acid hydrogels: natural, biodegradable tissue engineering scaffolds. Biotechnol Bioeng 82(5):578–589
37. Bella J, Eaton M, Brodsky B, Berman HM (1994) Crystal and molecular structure of a collagen-like peptide at 1.9A resolution. Science 266:75
38. Van der Rest WJ, Dublet B, Champliaud M (1990) Fibril-associated collagens. Biomaterials 11:28
39. Pins GD, Christiansen DL, Patel R, Silver FH (1997) Self-assembly of collagen fibers: influence of fibrillar arrangement and decorin on mechanical properties. Biophys J 73:2164–2172
40. Lee J, Macosko CW, Urry DW (2001) Elastomeric polypentapeptides cross-linked into matrices and fibers. Biomacromolecules 2:170–179
41. Vollrath F (1999) Biology of spider silk. Int J Biol Macromol 24:81–88
42. Hinman MB, Jones JA, Lewis RV (2000) Synthetic spider silk: a modular fiber. Trends Biotechnol 18:374–379
43. Tatham AS, Shewry PR (2000) Elastomeric proteins: biological roles, structures, and mechanisms. Trends Biochem Sci 25:567–571
44. Van Beek JD, Beaulieu L, Schafer H, Demura M, Asakura T, Meier BH (2000) Solid-state NMR determination of the secondary structure of Samia cynthia ricini silk. Nature 405(6790):1077–1079
45. Simmons AH, Michal CA, Jelinski LW (1996) Molecular orientation and two-component nature of crystalline fraction of spider dragline silk. Science 271(5245): 84–87
46. Hayashi CY, Lewis RV (2000) Molecular architecture and evolution of a modular spider silk protein gene. Science 287:1477–1479
47. Zhou CZ, Confalonieri F, Medina N, Zivanovic Y, Esnault C, Yang T, Jacquet M, Janin J, Duguet M, Perasso R, Li ZG (2000) Fine organization of Bombyx mori fibroin heavy chain gene. Nucleic Acids Res 28:2413–2419
48. Jin HJ, Kaplan DL (2003) Mechanism of silk processing in insects and spiders. Nature 410:541–548
49. Urry DW (1988) Entropic elastic processes in protein mechanisms. I. Elastic structure due to an inverse temperature transition and elasticity due to internal chain dynamics. J Protein Chem 7:1–34
50. Urry DW (1984) Protein elasticity based on coformation of sequential polypeptides – the biological elastic fiber. J Protein Chem 3:403–436
51. Tamburro AM, Guantieri V, Pandolfo L, Scopa A (1990) Synthetic fragments and analogues of elastin. II. Conformational studies. Biopolymers 29:855–870
52. Megret C, Lamure A, Pieraggi MT, Lacabanne C, Guantieri V, Tamburro AM (1993) Solid-state studies on synthetic fragments and analogues of elastin. Int J Biol Macromol 15:305–312

53. Debelle L, Tamburro AM (1999) Elastin: molecular description and function. Int J Biochem Cell Biol 31:261–272
54. Chilkoti A, Dreher MR, Meyer DR (2002) Design of thermally responsive, recombinant polypeptide carriers for targeted drug delivery. Adv Drug Deliv Rev 54:1093–1111
55. Wright ER, Conticello VP (2002) Self-assembly of block copolymers derived from elastin-mimetic polypeptide sequences. Adv Drug Deliv Rev 54:1057–1073
56. Yamaguchi I, Itoh S, Suzuki M, Sakane M, Osaka A, Tanaka J (2003) The chitosan prepared from crab tendon I: the characterization and mechanical properties. Biomaterials 24:2031–2036
57. Betre H, Setton LA, Meyer DE, Chilkoti A (2002) Characterization of a genetically engineered elastin-like polypeptide for cartilagenous tissue repair. Biomacromolecules 3:910–916
58. Miao M, Bellingham CM, Stahl RJ, Sitarz EF, Lane CJ, Keeley FW (2003) Sequence and structure determinants for the self-aggregation of recombinant polypeptides modeled after human elastin. J Biol Chem 278(49):48553–48562
59. Urry DW (1995) Elastic biomolecular machines. Sci Am 1:64–69
60. Bellingham CM, Lillie MA, Gosline JM, Wright GM, Starcher BC, Bailey AJ, Woodhouse KA, Keeley FW (2003) Recombinant human elastin polypeptides self-assemble into biomaterials with elastin-like properties. Biopolymers 70(4):445–455
61. Yang S, Leong K-F, Du Z, Chua C-K (2001) The design of scaffolds for use in tissue engineering. Tissue Eng 7(6):679–689
62. Lee KY, Rowley JA, Eiselt P, Moy EM, Bouhadir KH, Mooney DJ (2000) Controlling mechanical and swelling properties of alginate hydrogels independently by cross-linker type and cross-linking density. Macromolecules 33:4291–4294
63. LeRoux MA, Guilak F, Setton L (1999) Compressive and shear properties of alginate gel: Effects of sodium ions and alginate concentration. J Biomedial Mater Res 47:46–53
64. Eiselt P, Yeh J, Latvala RK, Shea LD, Mooney DJ (2000) Porous carriers for biomedical applications based on alginate hydrogels. Biomaterials 21:1921–1927
65. Shapiro L, Cohen S (1997) Novel alginate sponges for cell culture and transplantation. Biomaterials 18(8):583–590
66. Klock G, Pfeffermann A, Ryser C, Grohn P, Kuttler B, Hahn H-J, Zimmermann U (1997) Biocompatibility of mannuronic acid-rich alginate. Biomaterials 18:707–713
67. Zmora S, Glicklis R, Cohen S (2002) Tailoring the pore architecture in 3-D alginate scaffolds by controlling the freezing regime during fabrication. Biomaterials 23:4087–4094
68. Kuo CK, Ma PX (2001) Ionically crosslinked alginate hydrogels as scaffolds for tissue engineering: Part 1. Structure, gelation rate, and mechanical properties. Biomaterials 22:511–521
69. Hirano S, Midorikawa T (1998) Novel method for the preparation of N-acylchitosan fiber and N-acylchitosan-cellulose fiber. Biomaterials 19:293–297
70. Madihally SV, Matthew HW (1999) Porous chitosan scaffolds for tissue engineering. Biomaterials 20(12):1133–1142
71. Khor E, Lim LY (2003) Implantable applications of chitin and chitosan. Biomaterials 24:2339–2349
72. Chenite A, Chaput C, Wang D, Combes C, Buschmann MD, Hoemann CD, Leroux JC, Atkinson BL, Binette F, Selmani A (2000) Novel injectable neutral solutions of chitosan form biodegradable gels in situ. Biomaterials 21:2155–2161

73. Joshi HN, Topp EM (1992) Hydration in hyaluronic acid and its esters using differential scanning calorimetry. Int J Pharmaceut 80:213–225
74. Matthews JA, Wnek GE, Simpson DG, Bowlin GL (2002) Electrospinning of collagen nanofibers. Biomacromolecules 3:232–238
75. Sachlos E, Reis N, Ainsley C, Derby B, Czernuszka JT (2003) Novel collagen scaffolds with predefined internal morphology made by solid freeform fabrication. Biomaterials 24:1487–1497
76. Altman GH, Horan RL, Lu HH, Moreau J, Martin I, Richmond JC, Kaplan DL (2002) Silk matrix for tissue engineered anterior cruciate ligaments. Biomaterials 23:4131–4141
77. Lazaris A, Arcidiacono S, Huang Y, Zhou J-F, Duguay F, Chretien N, Welsh EA, Soares JW, Karatzas CN (2002) Spider silk fibers spun from soluble recombinant silk produced in mammalian cells. Science 295:472–476
78. Winkler S, Wilson D, Kaplan DL (2000) Controlling beta-sheet assembly in genetically engineering silk by enzymatic phosphorylation/dephosphorylation. Biochemistry 39(41):12739–12746
79. Winkler S, Szela S, Avtges P, Valluzi R, Kirschner DA, Kaplan D (1999) Designing recombinant spider silk proteins to control assembly. Int J Biol Macromol 24:265–270
80. Seidel A, Liivak O, Calve S, Adaska J, Ji G, Yang Z, Grubb D, Zax DB, Jelinski LW (2000) Regenerated spider silk: processing, properties, and structure. Macromolecules 33:775–780
81. Jin HJ, Fridrikh SV, Rutledge GC, Kaplan DL (2002) Electrospinning Bombyx mori silk with poly(ethylene)oxide. Biomacromolecules 3(6):1233–1239
82. Sofia S, McCarthy MB, Gronowicz G, Kaplan DL (2001) Functionalized silk-based biomaterials for bone formation. J Biomed Mater Res 54:139–148
83. Chen J, Altman GH, Karageorgiou V, Horan R, Collette A, Vollach V, Colabro T, Kaplan DL (2003) Human bone marrow stromal cell and ligament fibroblast responses on RGD-modified silk fibers. J Biomed Mater Res 67A(2):559–570
84. Vyavahare N, Ogle M, Schoen FJ, Levy RJ (1999) Elastin calcification and its prevention with aluminum chloride pretreatment. Am J Pathol 155:973–982
85. Daamen WF, Hafmans T, Veerkamp JH, van Kuppevelt TH (2001) Comparison of five procedures for the purification of insoluble elastin. Biomaterials 22:1997–2005
86. Meyer DE, Chilkoti A (2002) Genetically encoded synthesis of protein-based polymers with precisely defined molcular weight and sequence by recursive directional ligation: examples from the elastin-like polypeptide system. Biomacromolecules 3(2):357–367
87. Trabbic-Carlson K, Setton LA, Chilkoti A (2003) Swelling and mechanical behaviours of chemically cross-linked hydrogels of elastin-like polypeptides. Biomacromolecules 4(3):572–580
88. Lee CH, Singla A, Lee Y (2001) Biomedical applications of collagen. Int J Pharm 221:1–22
89. Kohle P, Kannan RM (2003) Improvement in ductility of chitosan through blending and copolymerization with PEG: FTIR investigation of molecular interactions. Biomacromolecules 4:173–180
90. Milella E, Brescia E, Massaro C, Ramires PA, Miglietta MR, Fiori V, Aversa P (2002) Physico-chemical properties and degradability of non-woven hyaluronan benzylic esters as tissue engineering scaffolds. Biomaterials 23:1053–1063
91. Gentleman E, Lay AN, Dickerson DA, Naumann EA, Livesay GA, Dee KC (2003) Mechanical characterization of collagen fibers and scaffold for tissue engineering. Biomaterials 24(21):3805–3813

92. Perez-Rigueiro J, Viney C, Llorca J, Elices M (2000) Mechanical properties of single-brin silkworm silk. J Appl Polym Sci 75:1270–1277
93. Cunniff J, Fossey S, Song J, Auerbach M, Kaplan DL, Eby R, Adams W, Vezzie D (1994) Mechanical and thermal properties of Nephila clavipes dragline silk. Polym Adv Technol 5:401–410
94. Lee KY, Bouhadir KH, Mooney DJ (2002) Evaluation of chain stiffness of partially oxidized polyguluronate. Biomacromolecules 3(6):1129–1134
95. Kim BS, Mooney DJ (2000) Scaffolds for engineering smooth muscle under cyclic mechanical strain conditions. J Biomech Eng 122(3):210–215
96. Urry DW (1999) Elastic molecular machines in metabolism and soft-tissue restoration. Trends Biotechnol 17:249–257
97. Elbjeirami WM, Yonter EO, Starcher BC, West JL (2003) Enhancing mechanical properties of tissue-engineered constructs via lysyl oxidase crosslinking activity. J Biomed Mater Res 66A(3):513–521
98. Anseth KS, Bowman CN, Brannon-Peppas L (1996) Mechanical properties of hydrogels and their experimental determination. Biomaterials 17(17):1647–1657
99. Tomihata K, Ikada Y (1997) In vitro and in vivo degradation of films of chitin and its deacetylated derivatives. Biomaterials 18:567–575
100. Avitabile T, Marano F, Castiglione F, Bucolo C, Cro M, Ambrosio L, Ferranto C, Reibaldi A (2001) Biocompatibility and biodegradation of intravitreal hyaluronan implant in rabbits. Biomaterials 22(3):195–200
101. Zhong SP, Compoccia D, Doherty PJ, Williams RL, Benedetti L, Williams DG (1994) Biodegradation of hyaluronic acid derivatives by hyaluronidase. Biomaterials 15(5):359–365
102. Benedetti L, Cortivo R, Berti T, Pea F, Mazzo M, Moras M, Abatangelo G (1993) Biocompatibility and biodegradation of different hyaluronan derivates (HYAFF) implanted in rats. Biomaterials 14(15):1135–1139
103. Campoccia D, Hunt JA, Doherty PJ, Zhong SP, O'Regan M, Benedetti L, Williams DF (1996) Quantitative assessment of the tissue response to films of hyaluronan esters. Biomaterials 17(10):963–975
104. Mi F-L, Wu Y-B, Shyu S-S, Schoung J-Y, Huang Y-B, Tsai Y-H, Hao J-Y (2001) Control of wound infections using a bilayer chitosan wound dressing with sustainable antibiotic delivery. J Biomed Mater Res 59:438–449
105. Van Wachem PB, Zeeman R, Dijkstra PJ, Feijen J, Hendriks M, Cahalan PT, van Luyn MJA (1999) Characterization and biocompatibility of epoxy-crosslinked dermal sheep collagens. J Biomed Mater Res 47(2):270–277
106. Greenwald D, Shumway S, Albear P, Gottlieb L (1994) Mechanical comparison of 10 suture materials before and after in vivo incubation. J Surg Res 56:372–377
107. Urry DW, Pattanaik A, Xu J, Woods TC, McPherson DT, Parker TM (1998) Elastic protein based polymers in soft tissue augmentation and generation. J Biomater Sci Polym Ed 9(10):1015–1048
108. Bouhadir KH, Lee KY, Alsberg E, Damm KL, Anderson KW, Mooney DJ (2001) Degradation of partially oxidized alginate and its potential application for tissue engineering. Biotechnol Prog 17:945–950
109. Alsberg E, Kong HJ, Hirano Y, Smith MK, Albeiruti A, Mooney DJ (2003) Regulating bone formation via controlled scaffold degradation. J Dental Res 82(11):903–908
110. Harriger MD, Supp AP, Warden GD, Boyce ST (1997) Glutaraldehyde crosslinking of collagen substrates inhibits degradation in skin substitutes grafted to athymic mice. J Biomed Mater Res 35(2):137–145

111. Altman GH, Diaz F, Jakuba C, Calabro T, Horan RL, Chen J, Lu H, Richmon D, Kaplan DL (2003) Silk-based biomaterials. Biomaterials 24:401–416

112. Vyavahare N, Jones PF, Tallapragada S, Levy RJ (2000) Inhibition of matrix metalloprotease activity attenuates tenascin-C production and calcification of implanted purified elastin in rats. Am J Pathol 157(3):885–893

113. VandeVord PJ, Matthew HWT, De Silva SP, Mayton L, Wu B, Wooley PH (2001) Evaluation of the biocompatibility of a chitosan scaffold in mice. J Biomed Mater Res 59:585–590

114. Setzen G, Williams EF (1997) Tissue response to suture materials implanted subcutaneously in a rabbit model. Plast Reconstruct Surg 100:1788–1795

115. Dahlke H, Dociu N, Thurau K (1980) Thrombogenicity of different suture materials as revealed by scanning electron microscopy. J Biomed Mater Res 14:251–268

116. Lee KY, Ha WS, Park WH (1995) Blood compatibility and biodegradability of partially N-acylated chitosan derivatives. Biomaterials 16:1211–1216

117. Muzzarelli R, Badassarre V, Conti F, Ferrara P, Biagini G, Gazzanelli G, Vasi V (1988) Biological activity of chitosan: ultrastructural study. Biomaterials 9:247–252

118. Otterlei M, Ostgaard K, Skjaek-Braek G, Smidsrod O, Soon-Shiong P, Espevik T (1991) Induction of cytokine production from human monocytes stimulated with alginate. J Immunother 10:286–291

119. Becker TA, Kipke DR, Brandon T (2001) Calcium alginate gel: a biocompatible and mechanically stable polymer for endovascular implantation. J Biomed Mater Res 54(1):76–86

120. Klock G, Frank H, Houben R (1994) Production of purified alginates suitable for use in immunoisolated transplantation. Appl Microbiol Biotechnol 40:638–643

121. Campoccia D, Hunt JA, Doherty PJ, Zhong SP, Callegaro L, Benedetti L, Williams DF (1993) Human neutrophil chemokinesis and polarization induced by hyaluronic acid derivatives. Biomaterials 14(15):1135–1139

122. Peluso G, Petillo O, Ranieri M, Santin M, Ambrosio L, Calabro D, Avallone B, Balsamo G (1994) Chitosan-mediated stimulation of macrophage function. Biomaterials 15(15):1215–1220

123. Usami Y, Okamoto Y, Takayama T, Shigemasa Y, Minami S (1998) Chitin and Chitosan stimulate canine polymorphonuclear cells to release leukotriene B4 and prostaglandin E2. J Biomed Mater Res 42:517–522

124. Lee KY, Alsberg E, Mooney DJ (2001) Degradable and injectable poly(aldehyde guluronate) hydrogels for bone tissue engineering. J Biomedial Mater Res 56:228–233

125. Trasciatti S et al. (1998) In vitro effects of different formulations of bovine collagen on cultured human skin. Biomaterials 19(10):897–903

126. Uchio YU, Ochi M, Matsusaki M, Kurioka H, Katsube K (2000) Human chondrocyte differentiation and matrix synthesis cultured in Atellocollagen gel. J Biomed Mater Res 50(2):138–143

127. Koob TJ, Willis TA, Hernandez DJ (2001) Biocompatibility of NDGA-polymerized collagen fibers. I. Evaluation of cytotoxicity with tendon fibroblasts in vitro. J Biomed Mater Res 56(1):31–39

128. Chevallay B, Abdul-Malak N, Herbage D (2000) Mouse fibroblasts in long-term culture within collagen three-dimensional scaffolds: Influence of cross-linking with diphenylphosphorylazide on matrix reorganization, growth, and biosynthetic and proteolytic activities. J Biomed Mater Res 49:448–459

129. Olde Damink LLH, Dijkstra PJ, Van Luyn MJA (1996) Cross-linking of dermal sheep collagen using a water soluble carbodiimide. Biomaterials 17:765–773

130. Van Luyn MJA, Van Wachem PB, Dijkstra PJ, Olde Damink L, Feijen J (1995) Cal-cification of subcutaneously implanted collagens in relation to cytotoxicity, cellular interactions, and crosslinking. J Mater Sci Mater Med 6:288–296

131. Van Luyn MJA, Van Wachem PB, Olde Damink L (1992) Relations between in vitro cytotoxicity and crosslinked dermal sheep collagen. J Biomed Mater Res 26:1091–1110

132. Gough JE, Scotchford CA, Downes S (2002) Cytotoxicity of glutaraldehyde crosslinked collagen/poly(vinyl alcohol) films is by the mechanism of apoptosis. J Biomed Mater Res 61(1):121–130

133. Van Wachem PB, van Luyn MJA, Olde Damink LHH, Dijkstra PJ, Feijen J (1994) Bio-compatibility and tissue regenerating capacity of crosslinked dermal sheep collagen. J Biomed Mater Res 28:353–363

134. Postlethwait RW (1970) Long-term comparative study of non-absorbable sutures. Ann Surg 171:892–898

135. Wen CM, Ye ST, Zhou LX, Yu Y (1990) Silk-induced asthma in children: a report of 64 cases. Ann Allerg 65:375–378

136. Zaoming W, Codina R, Fernandez-Caldas E, Lockey RF (1996) Partial character-ization of the silk allergens in mulberry silk extract. J Inv Allerg Clin Immunol 6:237–241

137. Lee KY, Kong SJ, Park WH, Ha WS, Kwon IC (1998) Effect of surface properties on the antithrombogenicity of silk fibroin/S-carboxymethyl kerateine blend films. J Biomater Sci Polym Ed 9:905–914

138. Santin M, Motta A, Freddi G, Cannas M (1999) In vitro evaluation of the inflamma-tory poptential of the silk fibroin. J Biomed Mater Res 46:382–389

139. Panilaitis B, Altman GH, Chen J, Jin HJ, Karageorgiou V, Kaplan DL (2003) Macrophage responses to silk. Biomaterials 24(18):3079–3085

140. Nicol AJ, Gowda DC, Urry DW (1992) Cell adhesion and growth on synthetic elas-tomeric matrices containing Arg – Gly – Asp – Ser-3. J Biomed Mater Res 26:393–413

141. Zielinski BA, Aebischer P (1994) Chitosan as a matrix for mammalian cell encapsu-lation. Biomaterials 15(13):1049–1056

142. Lahiji A, Sohrabi A, Hungerford DS, Frondoza CG (2000) Chitosan supports the ex-pression of extracellular matrix proteins in human osteoblasts and chondrocytes. J Biomed Mater Res 51:586–595

143. Lu JX, Prudhommeaux F, Meunier A, Sedel L, Guillemin G (1999) Effects of chitosan on rat knee cartilages. Biomaterials 20(20):1937–1944

144. Chang CNC, Rowley JA, Tobias G, Genes NG, Roy AK, Mooney DJ, Vacanti CA, Bonassar LJ (2001) Injection molding of chondrocyte/alginate constructs in the shape of facial implants. J Biomed Mater Res 55:503–511

145. Glicklis R, Shapiro L, Agbaria R, Merchuk JC, Cohen S (2000) Hepatocyte behaviour within three-dimensional porous alginate scaffolds. Biotechnol Bioeng 67(3):345–353

146. Dar A, Shachar M, Leor J, Cohen S (2002) Cardiac tissue engineering: optimization of cardiac cell seeding and distribution in 3D porous alginate scaffolds. Biotechnol Bioeng 80(3):305–312

147. Chang SC, Tobias G, Roy AK, Vacanti CA, Bonassar LJ (2003) Tissue engineering of autologous cartilage for craniofacial reconstruction by injection molding. Plast Reconstruct Surg 112(3):793–799

148. Shu XZ, Ghosh K, Liu Y, Palumbo FS, Luo Y, Clark RA, Prestwich GD (2004) Attachment and spreading of fibroblasts on an RGD peptide-modified injectable hyaluronan hydrogel. J Biomed Mater Res 68A(2):365–375

149. Grigolo B, Roseti L, Fiorini M, Fini M, Giavaresi G, Aldini NN, Giardino R, Facchini A (2001) Transplantation of chondrocytes seeded on hyaluronan derivative (Hyaff®-11) into cartilage defects in rabbits. Biomaterials 22:2417–2424
150. Suzuki Y, Tanihara M, Suzuki K, Saitou A, Sufan W, Nishimura Y (2000) Alginate hydrogel linked with synthetic oligopeptide derived from BMP-2 allows ectopic osteoinduction in vivo. J Biomed Mater Res 50:405–409
151. Lisignoli G, Fini M, Giavaresi G, Aldini NN, Toneguzzi S, Facchini A (2002) Osteogenesis of large segmental radius defects enhanced by basic fibroblast growth factor activated bone marrow stromal cells grown on non-woven hyaluronic acid-based polymer scaffold. Biomaterials 23:1043–1051
152. Rocha LB, Goissis G, Rossi MA (2002) Biocompatibility of anionic collagen matrix as scaffold for bone healing. Biomaterials 23:449–456
153. Altman GH, Horan RL, Martin I, Farhadi J, Stark PR, Volloch V, Richmond JC, Vunjak-Novakovic G, Kaplan DL (2002) Cell differentiation by mechanical stress. FASEB J 16(2):270–272
154. Altman GH, Horan RL, Martin I, Farhadi J, Stark PRH, Volloch V, Richmond JC, Vunjak-Novakovic G, Kaplan DL (2002) Cell differentiation by mechanical stress. FASEB J 16:270–272
155. Girotto D, Urbani, Brun P, Renier D, Barbucci R, Abatangelo G (2003) Tissue-specific gene expression in chondrocytes grown on three-dimensional hyaluronic acid scaffolds. Biomaterials 24:3265–3275
156. Saldanha V, Grande DA (2000) Extracellular matrix protein gene expression of bovine chondrocytes cultured on resorbable scaffolds. Biomaterials 21(23):2427–2431
157. Schuman L, Buma P, Verseleyen D, deMan B, van der Kraan PM, van den Berg WB, Hommoinga GN (1995) Chondrocyte behaviour within different types of collagen gels in vitro. Biomaterials 16:809–814
158. Nehrer S, Breinan HA, Ramappa A, Shortkroff S, Young G, Minas T, Sledge CB, Yannas IV, Spector M (1997) Canine chondrocytes seeded in type I and type II collagen implants investigated in vitro. J Biomed Mater Res 38(2):95–104
159. Nehrer S, Breinan HA, Ramappa R, Hsu H-P, Minas T, Shortkroff S, Sledge CB, Yannas IV, Spector M (1998) Chondrocyte-seeded collagen matrices impanted in a chondral defect in a canine model. Biomaterials 19(24):2313–2328
160. Huynh T, Abraham G, Murray J, Brockbank K, Hagen P-O, Sullivan S (1999) Remodeling of an acellular collagen graft into a physiologically responsive neovessel. Nat Biotechnol 17:1083–1086
161. Nehrer S, Breinan HA, Ramappa A, Young G, Shortkroff S, Louie L, Sledge CB, Yannas IV, Spector M (1997) Matrix collagen type and pore size influence on the behaviour of seeded canine chondrocytes. Biomaterials 18(11):769–776
162. Lee CR, Grodzinsky AJ, Spector M (2003) Biosynthetic response of passaged chondrocytes in a type II collagen scaffold to mechanical compression. J Biomed Mater Res 64A:560–569
163. Altman GH, Horan RL, Martin I, Farhadi J, Stark PR, Volloch V, Richmond JC, Vunjak-Novakovic G, Kaplan DL (2002) Cell differentiation by mechanical stress. FASEB J 16(2):270–272
164. Radice M, Brun P, Cortivo R, Scapinelli R, Battaliard C, Abatangelo G (2000) Hyaluronan-based biopolymers as delivery vehicles for bone marrow-derived mesenchymal progenitors. J Biomed Mater Res 50(2):101–109
165. Alsberg E, Anderson KW, Albeiruti A, Fransceschi RT, Mooney DJ (2001) Cell-interactive alginate hydrogels for bone tissue engineering. J Dent Res 60:2025–2029

166. Rowley JA, Mooney DJ (2002) Alginate type and RGD density control myoblast phenotype. J Biomed Mater Res 60(2):217–223
167. Rowley JA, Madlambayan G, Mooney DJ (1999) Alginate hydrogels as synthetic extracellular matrix materials. Biomaterials 20:45–53
168. Kale S, Biermann S, Edwards C, Tarnowski C, Morris M, Long MW (2000) Three-dimensional cellular development is essential for ex vivo formation of human bone. Nat Biotechnol 18(9):954–958
169. Cochran DL, Jones AA, Lilly LC, Fiorellini JP, Howell H (2000) Evaluation of recombinant human bone morphogenetic protein-2 in oral applications including the use of endosseous implants: 3 year results of a pilot study in humans. J Periodontol 71(8):1241–1257
170. Dvir-Ginsberg M, Gamlieli-Bonshtein I, Agbaria R, Cohen S (2003) Liver tissue engineering within alginate scaffolds: effects of cell-seeding density on hepatocyte viability, morphology, and function. Tissue Eng 9(4):757–766
171. Marler JJ, Guha A, Rowley J, Koka R, Mooney D, Upton J, Vacanti JP (2000) Soft-tissue augmentation with injectable alginate and syngeneic fibroblasts. Plast Reconstruct Surg 105(6):2049–2058
172. Perets A, Baruch Y, Weisbuch F, Shoshany G, Neufeld G, Cohen S (2003) Enhancing the vascularization of three-dimensional porous alginate scaffolds by incorporating controlled release basic fibroblast growth factor microspheres. J Biomed Mater Res 65A:489–497
173. Urry DW, Pattanaik A (1997) Elastic protein based materials in tissue reconstruction. Ann NY Acad Sci 831:32–46
174. Alsberg E, Anderson KW, Albeiruti A, Rowley JA, Mooney DJ (2002) Engineering growing tissues. Proc Natl Acad Sci 99(19):12025–12030
175. Wong Po Foo C, Kaplan DL (2002) Genetic engineering of fibrous proteins: spider dragline silk and collagen. Adv Drug Deliv Rev 54:1131–1143
176. Toman PD, Chisholm G, McMullin H, Gieren LM, Olsen DR, Kovach RJ, Leigh SD, Fong BE, Chang R, Daniels GA, Berg RA, Hitzemann RA (2000) Production of recombinant human type I procollagen trimers using a four-gene expression system in the yeast Saccharomyces cerevisiae. J Biol Chem 275:23303–23309
177. O'Brien JP, Fahnestock SR, Termonia Y, Gardner KH (1998) Nylons from nature: Synthetic analogs to spider silk. Adv Mater 10(15):85–95
178. Fahnestock SR, Bedzyk LA (1997) Production of synthetic spider dragline silk protein in Pichia pastoris. Appl Microbiol Biotechnol 47(1):33–39
179. Zhou Y, Wu S, Conticello VP (2001) Genetically directed synthesis and spectroscopic analysis of a protein polymer derived from a flagelliform silk sequence. Biomacromolecules 2:111–115

Author Index Volumes 101–102

Subject Index

Printing: Krips bv, Meppel
Binding: Stürtz, Würzburg